Cambridge El

Elements in Earth System
edited by
Frank Biermann
Utrecht University
Aarti Gupta
Wageningen University
Michael Mason
London School of Economics and Political Science

THE POLITICS OF DEEP TIME

Frederic Hanusch
*Justus Liebig University Giessen
and THE NEW INSTITUTE*

CAMBRIDGE UNIVERSITY PRESS

CAMBRIDGE UNIVERSITY PRESS

Shaftesbury Road, Cambridge CB2 8EA, United Kingdom

One Liberty Plaza, 20th Floor, New York, NY 10006, USA

477 Williamstown Road, Port Melbourne, VIC 3207, Australia

314–321, 3rd Floor, Plot 3, Splendor Forum, Jasola District Centre, New Delhi – 110025, India

103 Penang Road, #05–06/07, Visioncrest Commercial, Singapore 238467

Cambridge University Press is part of Cambridge University Press & Assessment, a department of the University of Cambridge.

We share the University's mission to contribute to society through the pursuit of education, learning and research at the highest international levels of excellence.

www.cambridge.org
Information on this title: www.cambridge.org/9781009468176

DOI: 10.1017/9781108936606

© Frederic Hanusch 2023

This work is in copyright. It is subject to statutory exceptions and to the provisions of relevant licensing agreements; with the exception of the Creative Commons version the link for which is provided below, no reproduction of any part of this work may take place without the written permission of Cambridge University Press & Assessment

An online version of this work is published at doi.org/10.1017/9781108936606 under a Creative Commons Open Access license CC-BY-NC-ND 4.0 which permits re-use, distribution and reproduction in any medium for non-commercial purposes providing appropriate credit to the original work is given. You may not distribute derivative works without permission. To view a copy of this license, visit https://creativecommons.org/licenses/by-nc-nd/4.0

All versions of this work may contain content reproduced under license from third parties.

Permission to reproduce this third-party content must be obtained from these third-parties directly.

When citing this work, please include a reference to the DOI 10.1017/9781108936606

First published 2023

A catalogue record for this publication is available from the British Library

ISBN 978-1-009-46817-6 Hardback
ISBN 978-1-108-94702-2 Paperback
ISSN 2631-7818 (online)
ISSN 2631-780X (print)

Cambridge University Press & Assessment has no responsibility for the persistence or accuracy of URLs for external or third-party internet websites referred to in this publication and does not guarantee that any content on such websites is, or will remain, accurate or appropriate.

The Politics of Deep Time

Elements in Earth System Governance

DOI: 10.1017/9781108936606
First published online: November 2023

Frederic Hanusch
Justus Liebig University Giessen and THE NEW INSTITUTE

Author for correspondence: Frederic Hanusch,
frederic.hanusch@planet.uni-giessen.de

Abstract: Human societies increasingly interact with processes on a geological or even cosmic timescale. Despite this recognition, we still lack a basic understanding of these interconnections and how they translate into politics. This Element provides an exploration and systematization of "the politics of deep time" as a novel lens of planetary politics in three steps. First, it demonstrates why deep-time interactions render the politics of deep time essential; second, it asks how deep time should be politicized; and third, it explicates the politics of deep time by examining representative cases. The Element also formulates a conceptual framework to open up possibilities for alliances that seek to better understand and realize the politics of deep time, pioneering a debate on how planetary temporalities can be politically institutionalized. This title is also available as Open Access on Cambridge Core.

This Element also has a video abstract: www.cambridge.org/Hanusch

Keywords: deep time, Earth system governance, long-term governance, democracy, geopolitics

© Frederic Hanusch 2023

ISBNs: 9781009468176 (HB), 9781108947022 (PB), 9781108936606 (OC)
ISSNs: 2631-7818 (online), 2631-780X (print)

Contents

1 Deep Time as a Novel Lens of Planetary Politics ... 1

2 The Why: Deep-Time Interactions ... 4

3 The How: Politicization of Deep Time ... 24

4 The What: Explicating the Politics of Deep Time ... 35

5 A Conceptual Framework of the Politics of Deep Time ... 72

References ... 78

1 Deep Time as a Novel Lens of Planetary Politics

When viewed in deep time, things come alive that seemed inert. New responsibilities declare themselves. A conviviality of being leaps to mind and eye. The world becomes eerily various and vibrant again. Ice breathes. Rock has tides. Mountains ebb and flow. Stone pulses. We live on a restless Earth.

(Macfarlane, 2019, pp. 15–16)

This Element investigates the politics of deep time as the realm in which societies interact with processes on a geological or even cosmic timescale. The aim is to examine deep-time interactions in order to provide a rationale for and conceptualization of the politics of deep time. As of now, the temporal depth of human actions conflicts with the short-termism of current political systems and institutions, which remain dominated by the election cycles of just a few years or policy programs of a few decades at best. Yet, the Anthropocene as an ongoing planetary event has brought to the foreground deep-time interconnections of human agency with the Earth system and, to an even greater extent, has highlighted that the Earth, while on very long timescales, is a restlessly changing and always provisional planet irrespective of human influence (Bauer et al., 2021; Gordon, 2021; Macfarlane, 2019). Despite this recognition of strong temporal interdependencies, we still lack a basic understanding of how societies can politically handle the interconnections between several decades, centuries, millennia, or eons, as well as the potential generations to come. In line with a previous call for "deep-time organizations" that exist over long periods of time to address deep-time challenges, this Element argues for the internationally coordinated establishment of a "deep-time observatory" (Hanusch & Biermann, 2020). The goal of the observatory would be to compile an inventory of deep-time interactions, in order to develop an evidence-based foundation for the politics of deep time as a core pillar, keeping the planet habitable and enabling the autonomy of future generations in the long run.

Within this introduction, I aim for a concise approximation of this politically and rather unfamiliar realm of deep time. I aim to show why it is distinct from and yet in most cases incorporates similar notions as other political concepts, such as the time policies for sustainability politics, politics of future generations, or long-term governance (Boston, 2016; Reisch, 2015; Underdal, 2010). Futures that referred to children's children in a sustainability context are replaced by Earth time periods, in which it is not even clear whether and in what form human societies will experience these (Horn, 2017). The politics of deep time are thus related to yet are distinct from politics concerned with a handful of generations in the past or future. Drawing from the spatial notion of multilevel governance, one has to develop the notion of a "multitemporal governance," this has only been conceptualized for the short-term level of election periods and the medium- to long-term level of certain future generations; the level of processes taking place within cosmic timescales, however, has to date been omitted. Therewith, I contribute to a wider paradigm shift in the global environmental politics research in the Anthropocene (Biermann, 2021).

Deep time is a realm far beyond human existence yet entangled with it. To better understand the character and scales of deep time, I start here with a brief analogy with everyday life, namely, food. First, imagine buying food for the next seven days. One would probably select certain foods that must be eaten within the first few days, such as lettuce, but one would also purchase items that can be cooked in a week from now, such as potatoes. Next, consider the seven-generation principle of the Haudenosaunee, according to which decisions today should benefit seven generations in the future. To guarantee that the seventh generation in the future can enjoy food, one would have to plant olive trees, so that future generations can also enjoy tasty food and make dining tables from the wood of the olive trees. Finally, try to imagine how food relates to the next seven geological epochs. The Cenozoic era, spanning approximately 66 million years, consists of seven geological epochs, namely, the Paleocene, Eocene, Oligocene, Miocene, Pliocene, Pleistocene, and Holocene. Within this timeframe, the Chicxulub asteroid impact occurred around 66 million years ago, leading to the extinction of non-avian dinosaurs. Thinking about our relationship with food that operates within these vast timeframes means ensuring that the genetic diversity of crops, which has been forming over millions of years, does not become extinct and allows the regeneration of biogeochemical soil flows that enable respective plantation. Within the last 2.5 billion years, for example, no other force had a greater impact on the nitrogen cycle than humans, largely due to nitrogen fertilizers used in agriculture (Canfield et al., 2010). Yet, we overlook political practices and institutions that are capable of dealing with these kinds of timeframes, timeframes so vast that numbers lose meaning. This encapsulates the politics of deep time.

While such relations with Earth system processes, which form over cosmic timescales, may sound fairly distant and technical at first glance, they are vastly politicized and drive world politics. When US president Trump, for example, claimed "OIL (ENERGY) IS BACK!!!!" (Trump, 2020), a "Great Again" retrotopia of a romanticized past based on fossil-fuels, formed ca. 286 to 360 million years ago, became a powerful future image of reactionary movements around the world (Hanusch & Meisch, 2022). This indicates that changes in civilizations are profoundly interrelated with changes in their conceptions of time. While several other disciplines have started investigating these interrelations, such as philosophy (Landa, 2000), history (Chakrabarty, 2009), or educational sciences (Zen, 2001) among others, an explicit treatment of the politics of deep time is overdue.

When comparing the politics of deep time to related temporal concepts in social science research, its distinct character can be further distilled. First, political time is concerned with diverse temporal understandings that are determined by political institutions and actors within the political process itself, such as the timing of decision-making or regime change (Goetz, 2019). Consequently, long-term politics refers to the long-term problems and policies over years, decades, or centuries, but rarely millennia or more (Siebenhüner et al., 2013). In a similar vein, anticipation studies, such as research on emerging technologies, are a methodological approach aiming to understand possible future trajectories and to act accordingly, mostly within a maximum time range of decades (Poli, 2017). Studies concerning the politics of future generations overwhelmingly focus on a few generations into the future (Boston, 2016; González-Ricoy & Gosseries, 2016), with the exception of some concepts proposing timeless trusteeship ideas (Thompson, 2010; see Section 3.2). Similar to geohistory or "une histoire quasi immobile" and the notion of the "longue durée" (Braudel, 1966, p. 16), Big History or the Climate of History can be partly related to the politics of deep time, as it strives to integrate human history with the history of the universe, without exemplifying what this means in terms of politics (Chakrabarty, 2021; Christian, 2011). Timescapes are yet another and probably the most encompassing approach in social theory that comprise a cluster of various interacting temporal phenomena, ranging from timing, tempo, duration, sequence, and timeframes to modalities of past, present, and future, but have primarily served as a theoretical approach (Adam, 1998). While the politics of deep time incorporate a range of the above-mentioned and similar approaches – the generation of our great-grandchildren is, for example, a tiny part of this – an explicit and comprehensive treatment of the inhuman cosmic timescales from a political perspective is, to the best of my knowledge, lacking.

This Element proceeds as follows: in order to outline a conceptual framework of the politics of deep time (Section 5), I investigate *why* deep-time interactions make the politics of deep time essential (Section 2), *how* deep time is currently politicized (Section 3), and *what* concrete cases should be treated as politics of deep time (Section 4). I thus introduce the notion of the politics of deep time from scratch as both an analytical framework and a political necessity. After all, we must learn to talk temporally about time. Acquiring a temporal view is a demanding exercise, yet it allows for novel insights into the ever-changing relationships between humans and the planet.

2 The Why: Deep-Time Interactions

> Rocks are not nouns but verbs.
>
> (Bjornerud, 2018, p. 8)

This section develops the very basis of a new kind of politics. It starts by defining deep time as the realm in which societies interact with processes on a geological or even cosmic timescale. Thereafter, deep-time encounters are investigated to unveil deep-time interactions. Based on this, I identify normative objectives for the politics of deep time, namely, democracy and habitability.

2.1 Definition of Deep Time

I define deep time as the realm in which societies interact with processes on a geological or even cosmic timescale. Deep time is thus a relational concept that covers the interactions between societies and processes taking place within geological and cosmic times. Although often used interchangeably, particularly in a non-geological discourse, deep, geological, and cosmic time can be distinguished with regard to their definition (Burchfield, 1998; McPhee, 1981).

Cosmic time encompasses the timeframe since the Big Bang ca. 13.8 billion years ago until the ultimate fate of the universe, which will eventually manifest in the form of a Big Crunch in ca. 20 billion years, a Big Rip in ca. 50 billion years or a Big Chill in ca. one googol year. Cosmic time is thus a synonym for the age of the universe, whereas geological time refers to the timeframe of the Earth's existence, ranging from its formation 4.54 billion years ago to its absorption by the Sun in ca. 7.5 billion years. Geological time is thus synonymous with the age of planet Earth. Geological time is part of cosmic time, with "us as creatures of this earth, as beings that are constituted by a double temporality: rhythmically structured within and embedded in the rhythmic organisation of the cosmos" (Adam, 1998, p. 13; see Figure 1).

The discovery of these vast amounts of time relating to the existence of the Earth and the Universe was in some cultures, at least in the Christian dominated parts of the world, preceded by a much more anthropocentric interpretation of the beginning of everything. A prime example is the assumption that the Earth is no older than a few thousand years (see Figure 2).

The proposition of a much older Earth, based on geological, rather than religious, timescales, dates back to at least the eleventh century to two polymaths, namely, Ibn Sina during the Islamic Golden Age and Shen Kuo during the Song dynasty. Going further back, Hinduism and Buddhism refer to the idea of "kalpa," which is similar to the notion of a cosmological eon. However, James Hutton is mostly referred to as the discoverer of geological time in which

The Politics of Deep Time

"we find no vestige of a beginning, no prospect of an end" (Hutton, 1788/2010, p. 304). In 1788, at Siccar Point on the east coast of Scotland, Hutton observed how two different rock types layered on top of each other, later known as "Hutton's unconformity," and concluded that the Earth's surface is the result of cyclic geological processes that are too slow to have taken place in biblical timeframes (see Figure 3). Stones thus may be perceived as a critter themselves; at least they have the potential to transfer people into nonhuman scales of time (Cohen, 2015; Reinert, 2016).

As his "Theory of the Earth" (1788/2010) was criticized as illogical and atheistic, for example, by other geologists such as Richard Kirwan, he published "An Investigation of the Principles of Knowledge and of the Progress of Reason, from Sense to Science and Philosophy" as a three-volume edition including his "Theory of the Earth" and additional material to justify his findings. John Playfair, a colleague who accompanied James Hutton at Siccar Point, described the recognition of the discovery a few years later as follows:

> We felt ourselves necessarily carried back to the time when the schistus on which we stood was yet at the bottom of the sea, and when the sandstone before us was only beginning to be deposited. ...Revolutions still more remote appeared in the distance of this extraordinary perspective. The mind seemed to grow giddy by looking so far into the abyss of time; and while we listened with earnestness and admiration to the philosopher who was now unfolding to us the order and series of these wonderful events, we became sensible how much farther reason may sometimes go than imagination can venture to follow. (Playfair, 1805, p. 73)

Geological time, in this vein, becomes materially accessible and visible through its presence in the here and now, "palpably present in rocks, landscapes, groundwater, glaciers, and ecosystems" (Bjornerud, 2018, p. 162), but remains in part invisible and inaccessible in the depths of the Earth or in the vastness of outer space (Chakrabarty, 2018; Szerszynski, 2017).

The discovery of geological time was not only a revolutionary moment in geology. It inspired the great theories of the sciences, including Charles Darwin's "The Origin of Species" (1859), as well as poets and novelists alike in terms of the implications for the missing justification of humans as the pride of creation (Buckland, 2013; Ziolkowski, 1990). As a consequence, it is argued that the discovery of geological time is next to the Copernican Revolution, Darwin's theory of evolution, and Freud's theorization of the subconscious, one of the four great revolutions which led to a decentering of human subjectivity: "What could be more comforting, what more convenient for human domination, than the traditional concept of a young earth, ruled by human will within days of its origin. How threatening, by contrast, the notion of an almost incomprehensible

NEOPROTEROZOIC
MESOPROTEROZOIC
PALEOPROTEROZOIC
.54 Ga CAMBRIAN EXPLOSION
.6 Ga OZONE LAYER
.64 Ga MARINOAN GLACIATION (SNOWBALL EARTH)
CRYOGENIAN ICE AGE
.7 Ga STURTIAN GLACIATION
.75 Ga FIRST ANIMALS
.9 Ga RODINIA SUPERCONTINENT
1.1 Ga FIRST PREY-PREDATOR
1.2 Ga SEXUAL REPRODUCTIVE COMMON ANCESTOR IN FOSSIL RECORD
1.3 Ga COPPER DEPOSIT FORMATIONS
1.4 Ga CYANOBACTERIA INVADES A EUKARYOTE AND BECOMES A CHLOROPLAST
1.5 Ga NUNA SUPERCONTINENT BREAKS
1.7 Ga VARIOUS OROGENIC EVENTS
1.8 Ga OLDEST SAND DUNES
1.9 Ga BACTERIAL VIRUSES
2 Ga MAJOR OROGENY (MOUNTAIN FORMATION)
2.1 Ga EARLIEST MULTICELLULAR ORGANISM
2.1-2.4 Ga GREAT OXYGENATION
2.3 Ga HURONIAN ICE AGE
2.4 Ga SUAVJARVI METEOR IMPACT
2.4 Ga END OF PURPLE EARTH (OXYGEN CRISIS)
2.5 Ga NUNA SUPERCONTINENT
2.7 Ga GLOBAL KOMATIITE ERUPTION
2.8 Ga UR SUPERCONTINENT
2.9 Ga PONGOLA GLACIATION
3 Ga RETINAL-BASED LIFE DOMINATES (PURPLE EARTH)
3.26 Ga LARGEST METEOR IMPACT
3.3 Ga JOHANNESBURG DOME
CYANOBACTERIA
CHLOROPLASTS
RED ALGAE
DINOFLAGELLATES
GLENOBOTRYDION
FUNGI
PARAMECIUM
AMOEBA
MELANOCYRILLIUM
SPONGE
ARTHROPODS
TRILOBITES
CRUSTACEANS
BACTERIOPHAGE
EUKARYOTES
STROMATOLITES
MOLECULAR CLOUD
PRIMAL SUPERNOVA
PRIMAL STAR
PROTOPLANETARY DISK
ANDROMEDA
DARK ENERGY OVERTAKES GRAVITY
MILKY WAY SPIRALS
SUN FORMATION
DARK AGES
CMB
SPIRAL GALAXIES
FIRST GENERATION STARS
INFLATION
POPULATION II STARS
QUASARS PEAK
GLOBULAR CLUSTERS
TAU CETI
MIRA
LALANDE 21185
ALPHA CENTAURI
GLIESE 581
BARNARD'S STAR
EARTH FORMATION
THEIA IMPACT
GAIA-ENCELADUS GALAXY MERGES WITH MILKY WAY
TERZAN 7
ARCTURUS
ALDEBARAN
LUYTEN'S STAR
ROSS 128
WATER
LATE HEAVY BOMBARDMENT

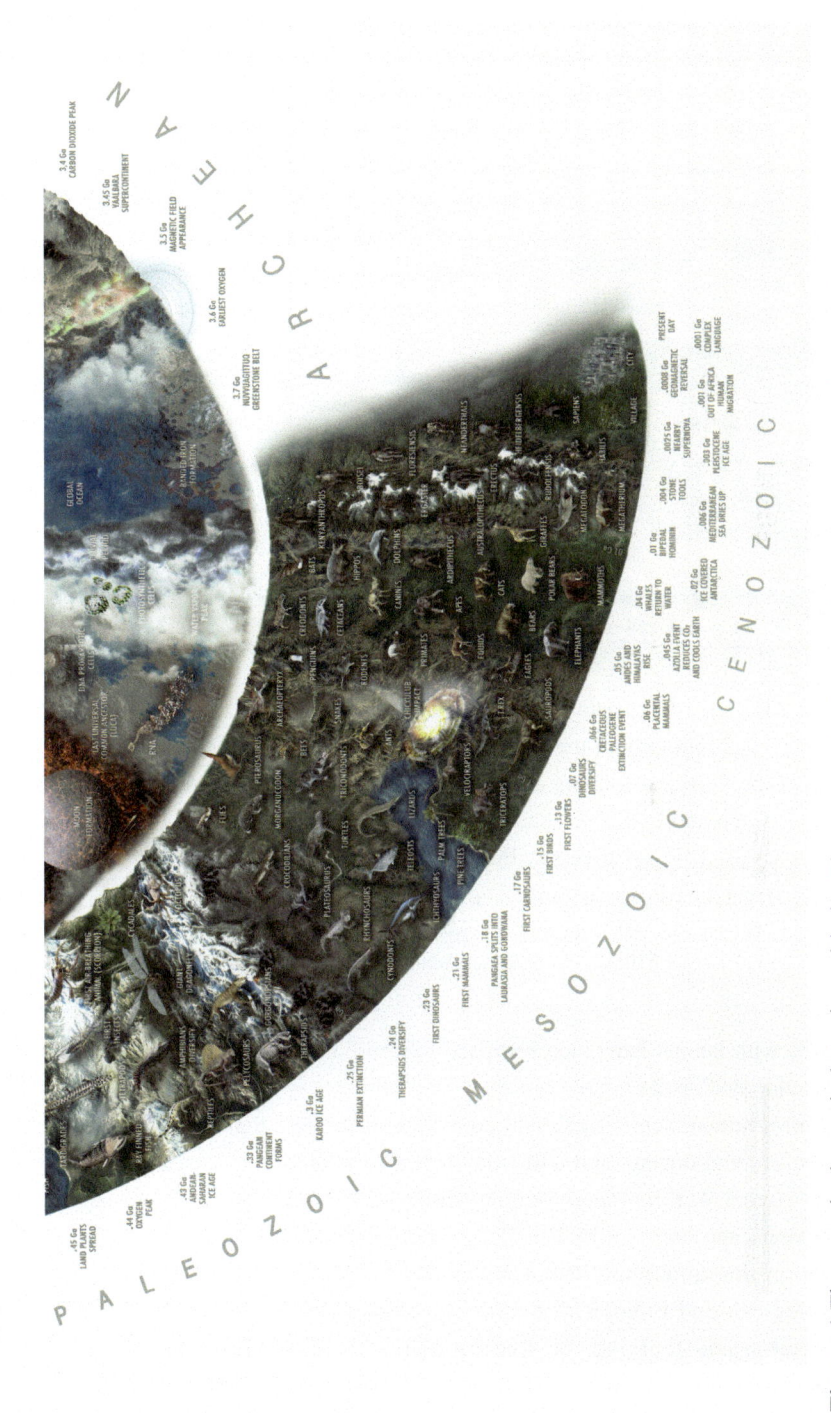

Figure 1 The cosmic time spiral starting with the Big Bang and a focus on the geological time of the Earth. A 90-degree stretch covers one billion years, with the most recent 90 degrees corresponding to only 500 million years.

Source: Reprinted with the permission of Pablo Carlos Budassi (2020), available at www.pablocarlosbudassi.com/2021/02/nature-timespiral.html

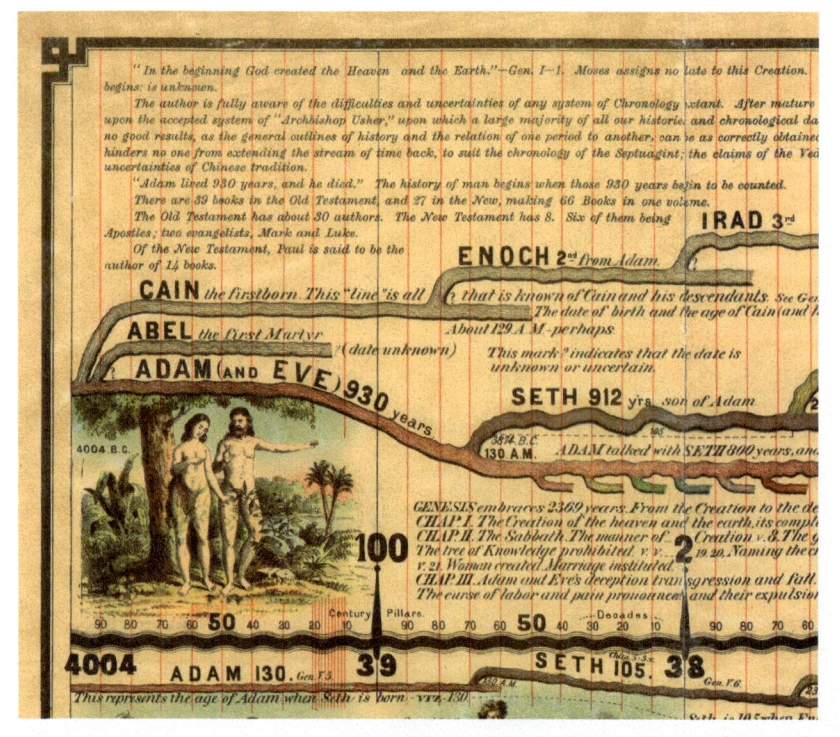

Figure 2 Details of Sebastian C. Adam's Synchronological Chart (1881) showing the chronology of the Earth according to James Ussher's (1650) Annals of the World, beginning at 6 pm on October 22, 4004 BC, with the creation of Adam and Eve. The chart was reproduced in Bibles by the Oxford University Press until 1910.

Source: © David Rumsey Map Collection, David Rumsey Map Center, Stanford Libraries, CC BY-NC-SA 3.0; available at www.davidrumsey.com/luna/servlet/detail/RUMSEY~8~1~226099~5505934:Composite–Adams–Synchronological-

immensity, with human habitation restricted to a millimicrosecond at the very end!" (Gould, 1987, p. 2).

Therefore, human societies should comprehend the fact that planet Earth existed before and during, and will exist after *Homo sapiens*. Consequently, the past covers a timeframe ranging from the first *Homo sapiens* ca. 315,000 years ago back to the Big Bang. A closer look at processes taking place within this timeframe demonstrates that the separation between cosmic, geological, and biological timeframes becomes blurred. Not only are there bacteria, and thus life, in Siberian soil which can repair their own DNA and survive for at least 500,000 years (Johnson et al., 2007), but also the origin of many minerals is related to biological processes, demonstrating

Figure 3 Hutton's unconformity at Siccar Point, Scotland. The lines illustrate the differing orientation of the strata between the two stone types from different ages.
Source: Image by Mike Brooks © Herefordshire & Worcestershire Earth Heritage Trust; available at https://deeptime.voyage/siccar-point/

the coevolution of life and minerals as a result of permanent interchange (Ehrlich, 1996). The time during the presence of *Homo sapiens* on planet Earth covers the timeframe of the existence of *Homo sapiens* from ca. 315,000 years ago toward a yet unknown point in the future, when a "homo noveau" might start to form, for example, by natural mutation, by separation of *Homo sapiens* on two planets, by genetic engineering or, in a transhuman manner, by integrating artificial intelligence devices, with the resulting relationship between both unfolding its agency. A future after the existence of *Homo sapiens* covers a timeframe ranging from the emergence of a homo noveau to the unknown fate of the universe. In the meantime, the Atlantic Ocean will close again and in ca. 250 million years, a new supercontinent, Pangea Proxima, will form; whether humans will exist then is unknown (Williams & Nield, 2007). Of course, such timeframes are hard to imagine, and yet, the very moment one reads these sentences, once also "used to be the unimaginable future" (Brand, 1999, p. 164).

Hutton's discovery thus brought about a bifurcation of the Earth's and human history, which the notion of the Anthropocene is reuniting again. This reunion becomes explicit by illuminating human interactions with processes taking place within geological timescales in the form of stratigraphic, biostratigraphic, and chemostratigraphic signatures and the physical stratigraphic signatures of the Technosphere (Northcott, 2015; Zalasiewicz et al., 2019).

How societies know about and perceive time on the one side, and how they are organized and govern themselves on the other side thus depend on each other. Societal relations regarding knowledge of the timeframes of the Earth's and the Universe's existence changed during the course of history and will likely be subject to change in the future. How societies perceive their relationship to these large timeframes fundamentally changes worldviews. Depending on societies' perception of the Earth's age, whether it is some thousand or some billion years old, their self-conception and politics differ. When it comes to such vast amounts of time, which may be measurable but are yet so unfamiliar that they can barely be comprehended, the way in which relationships with these are shaped become particularly important. This characterizes deep time.

In theoretical terms, the separation between societies and their understanding of time, in contrast to geological and cosmic time, aligns with the distinction made between an A- and B-series in time philosophy (McTaggart, 1908). The A-series refers to the subjective experience of time, where events are ordered dynamically according to their position in relation to the present moment, namely, past, present, or future. The Heraclitean character of the A-series can thus be closely aligned to societal time perceptions. The B-series refers to a rather objective and fixed sequence of events with either earlier or later than other events in t_1, t_2, t_3, and so on. The Parmidean character of the B-series thus aligns with the block universe underlying cosmic and geological time. Several concepts exist that potentially connect the A- and B-series, including the "specious present" (James, 1893; Kelly, 1882), "tensed facts" (Swinburne, 1990) and "time consciousness" (Dainton, 2023). However, these concepts usually refer to the individual and not societal level. Deep time is conceptualized here as a means of connecting the A- and B-series at a societal and political level. The societal experience of time is closely tied to the events that societies perceive taking place around them, while it is difficult to comprehend the vast timescales of geological or even cosmic time within societal experiences. Deep time can provide a bridge between the A-series and B-series of time. Deep time allows to connect the societal experiences of time with the reality of the universe, as deep time explicates the interactions between both.

In methodological terms, "Numbers do not seem to work with regard to deep time. Any number above a couple of thousand years – 50,000, 50 million – will with nearly equal effect awe the imagination" (McPhee, 1981, p. 21; see also Ginn et al., 2018, p. 214). An illustrative example of this is that 85 million years

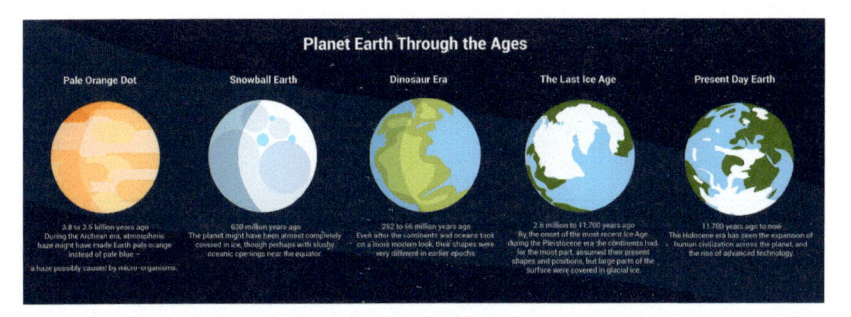

Figure 4 Planet Earth changed markedly many times in the past and will continue to do so in the future.

Source: © NASA/JPL-Caltech/Lizbeth B. De La Torre (2020); available at https://exoplanets.nasa.gov/resources/2245/planet-earth-through-the-ages/

lie between the appearance of the stegosaurus and the tyrannosaurus, and "only" 67 million years separate the appearance of the tyrannosaurus from the invention of the mobile phone. If, therefore, geological and cosmic time are the quantitative accounts of these huge timeframes, which are hardly accessible to the human mind, deep time can be understood as its qualitative account; we live within geological time and interact with it through deep-time interactions. My definition of deep time is thus a temporal explication and basis of politicization of the notion of "earthly multitudes" as the various connections diverse societies establish with the "planetary multiplicity" of an ever-changing planet (Clark & Szerszynski, 2021, pp. 171–172; see Figure 4).

More precisely, societal interactions through deep time with processes taking place within geological or cosmic timescales can include, for example, the way soils, formed over billions of years, enable or disable certain types of societal development, how currently produced nuclear waste will radiate a million years into the future, and the possible terraforming of Mars. The differentiated meanings and relationships of cosmic, geological, and deep time are shown in Figure 5.

Even though human evolutionary history is characterized by these interactions, most humans do not take notice of these; neither do societal institutions explicitly address them (Pahl et al., 2014).

2.2 Deep-Time Encounters

Having defined deep time, I now take a closer look at the kind of interactions between societies and processes taking place within geological timescales by asking: How has awareness of the importance of deep time as a policy-relevant dimension evolved? What characterizes deep-time encounters? Why are deep-time encounters politically relevant?

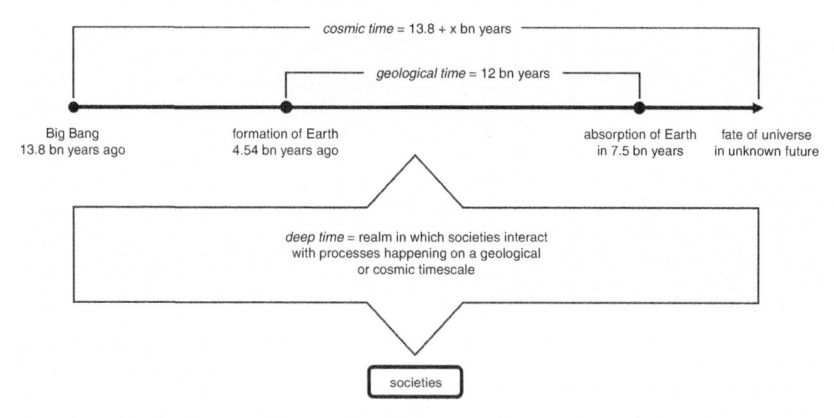

Figure 5 The differentiated meanings and relationships of cosmic, geologic, and deep time.

First, how has awareness of the importance of deep time as a policy-relevant dimension evolved? While the human impact on the long-lasting processes of the Earth's surface, through fire, extinctions, and deforestation, has been a traditional object of study in geomorphology for decades, the emergence of proto-Anthropocene planetary knowledge since the 1950s, in particular, later known as Earth system sciences, pointed to the entanglements between timescales and led to the respective integration into a geo-anthropological time (Goudie, 2020; Sörlin & Isberg, 2021; Figure 6). It became increasingly clear that humans are not only living on a planet, but are part of it, yet many are missing planetary knowledge that could form the basis for respective action (Hanusch et al., 2021). Moreover, such knowledge and the considerable timescales to which it relates were for a long time and are still partially denied and declared insignificant by large parts of the humanities and social sciences:

> How could it make any kind of sense to insert into the totality of the evolutionary process of the universe this tiny portion of a timespan that is illuminated by the light of tradition? Yet, what imposes itself upon us when it comes to broadening the historical horizons is precisely to stop thinking of this gigantic framework into which the little bit of human destiny called world history almost disappears. (Gadamer, 1988/2016, p. 27)

This is changing, as a distinct deep-time perspective is seen as fundamental to understanding human life itself: "To isolate life from these geological flows is to distort our understanding of society and of humanity. Yet all too often the focus upon an inflated present abducts contemporary activity from the geological duration needed to fully understand its significance – an extraction of the contemporary moment from deep time that threatens rupture" (Irvine, 2020, pp. 2–3). A respective thought experiment asks: What if humans lived for

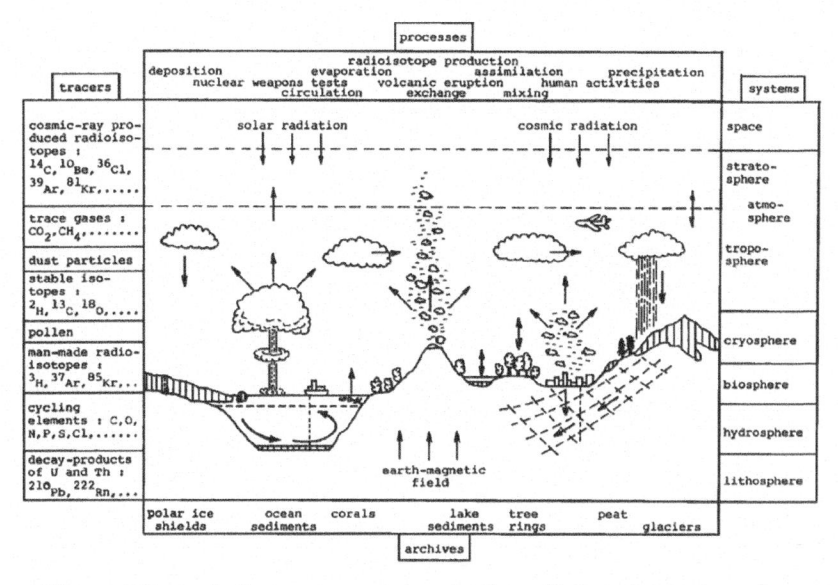

Figure 6 Proto-Anthropocene synchronization efforts to integrate various timescales into a geo-anthropological timeframe.

Source: Oeschger (1985, p. 10) © American Geophysical Union (AGU)

20,000 years and could see mountains moving or seas rising? (Chakrabarty, 2021, p. 191). They would not see rocks or deserts as part of a background landscape but part of an ever-changing planet: "The body of the earth is in motion, and our bodies within it" (Irvine, 2020, p. 189).

With all the knowledge regarding the temporalities of the Earth's systems, the formulation of the notion of the Anthropocene has to be viewed as an epistemic event that also called the humanities and social sciences to action (Bai et al., 2016; Bonneuil & Fressoz, 2013/2016). Not only has it diminished the separation between nature and culture, it has also brought the understandings of time implied in each of the concepts closer together. The Anthropocene opened an entry point into deep time that brought the distant past and the far away future into the present. This does not only include (potential) human influences – while making a "human planet" – comparable to a meteor that ended the Cretaceous 66 million years ago, but also an understanding of "how the forces of strata and Earth system work through us" (Clark & Szerszynski, 2021, p. 60; see also Chakrabarty, 2021, p. 8; Lewis & Maslin, 2018; Zalasiewicz & Kunkel, 2017; see Figure 7). This marked a counterpoint to the marginalization of the human by the vast geological timescales. During the Anthropocene, diverse societies gained agency to different degrees. This agency arises and operates on short and very long timescales, and brings about ruptures as well as steady processes that often surpass any measure and phenomenological ground of the human (Chakrabarty, 2021, p. 190; Clark & Szerszynski, 2021, p. 20; Hamilton, 2016). The role of humans became ambiguous; not only are they influenced by processes taking place within timescales of billions of years, but they are also, in part, capable of influencing, although not controlling, these processes. This influence can be used by some as a political force with the unifying notion of the Anthropocene as a "humanity-as-strata," readable after human extinction (Yusoff, 2016, p. 9, 2018b). This invokes more pluralistic notions, such as the Chthulucene, that places emphasis on the temporal plurality of permanently ongoing multispecies rearrangements (Colebrook, 2017, p. 10; Haraway, 2015).

Second, what characterizes deep-time encounters? These redefine agency in a twofold and reciprocal manner: on the one side, human societies have agency to shape processes taking place within geological and cosmic timescales; and on the other side, these processes themselves exert nonhuman agency influencing human societies. This is what I call deep-time interactions (see Figure 5).

Figure 7 Humans are in their daily life surrounded by and part of manifold deep-time interactions. Often these go unnoticed: in this picture people enjoy their time at Itzurun beach in the Spanish Basque Country, where the Earth's strata are only perceived as a backdrop for typical beach activities.

Source: © Dosfotos/Axiom (Design Pics Inc)/Alamy (2011), available at www.alamy.com/EYDB3C

Accordingly, deep-time interactions are always bidirectional. This means that in terms of the past, today's societies are influenced by processes that took place within a cosmic timescale, hundreds of thousands or even millions and billions of years ago; and, second, that societies influence these processes. Regarding the latter, this is not about time travel, but understanding, on the one side, which kind of memory the Earth as a planet practices and, on the other side, the way societies relate to these processes, for example, by declaring a geological site or a landscape a natural World Heritage Site, thus attaching value to a place that formed a very long time ago, which should not be altered or from which resources should not be extracted (Graeber & Wengrow, 2021; Szerszynski, 2019). In terms of the present, this means that societies of *Homo sapiens* are influenced by events in the here and now, caused by processes taking place within a cosmic timescale and that societies influence these processes. They may interact with resources that developed in the distant past of the Earth's history, using resources and thus potentially erasing parts of the Earth's memory, making it disappear. In addition, regarding the future, this means that today's societies are influenced by processes that are likely to take place within a cosmic timescale in the distant future and, second, that societies influence these processes. Once again, this does not relate to time travel, but whether and how societies react to future events predicted by evidence-based models, such as giant solar flares hitting Earth, and which tools are being invented to deal with future events taking place within cosmic timescales, ranging from technologies to cultural practices enabled by respective political arrangements.

The term "bidirectionality" thus refers to the reciprocal and interactive nature of the relationships between human societies and processes operating within geological or cosmic timescales. It recognizes that human activities and decisions can have an impact on these long-term processes, while acknowledging that the outcomes of these processes can, in turn, affect human societies. Bidirectionality emphasizes the interconnectedness and interdependence between human actions and the larger planetary or cosmic systems. It emphasizes that human societies are not isolated entities, but part of a complex web of interactions where deep-time interactions can be mutually influenced and coevolve between human societies and processes within geologic or cosmic timescales. Accordingly, bidirectionality is fundamentally different from unilateral causality, taking into account the construction of deep-time images and possibilities, their emergence, contextual embedding, and path dependencies (Polak, 1973). Geological processes, for example, erosion, sedimentation, and tectonic movements, shape the Earth's landforms over long periods of time. These processes, in turn, influence the development and change of cultural landscapes. The formation of valleys, mountains, and coastlines can affect settlement patterns, transportation networks, and the cultural practices of local communities, while at the same time, human societies rebuild these landscapes, even creating new islands (Sreekumar & Hassan, 2020).

One key characteristic of the twofold relationship of deep-time interactions is that nonhuman agency enables human life, yet it remains indifferent to it (Colebrook, 2017, p. 3). This agency is not only executed through geological material, such as soil or events like volcanic eruptions. The geological and biological are deeply intertwined in the Earth's history; bacteria which, as the sole inhabitants of Earth for a timespan of three billion years, laid the foundation for more complex and thus human life. They drove chemical reactions, transforming a previously purely geological planet and inventing the biosphere (Blaser, 2014); for this reason, "animals figure disproportionately in the maintenance of the modern Earth System, not least because they invented it" (Butterfield, 2011, p. 86). Being an active force that transforms the Earth's systems is thus not even exclusive to human life, it has been done before, albeit in a direction that in most cases increased diversity instead of diminishing it: "The first cell never died, it bifurcated in time, generated all biology, all of us, and all technology" (Walker [@Sara_Imari], 2022). Applying a deep-time lens goes beyond the very basics of what life is: it transforms the definition of life and living creatures to reflect the notion of holobionts, which emphasizes the importance of symbiosis and even symbiogenesis for the formation of life, instead of placing individuals first (Gilbert, 2014; Ginn et al., 2018, pp. 214–215). As I will discuss later, this also has consequences for the meaning of biodiversity loss beyond the conventional definitions of the loss of certain distinct species in the here and now. Life forms through genealogies, it preserves the past and binds time, it is "expanding complexity and creating new problems for itself" (Margulis & Sagan, 2000, p. 86).

Those primary forces – underlying but also including life itself – operating within geological timescales impacted humans much more directly than contemporary societies. Their impact, for example, through glacial action or shifts in vegetation and fauna, was not mitigated by modern infrastructure, such as larger cities or state organizations. This is why "the paleoenvironment and human response to it may be considered as two cogs within a well-calibrated precision instrument, where even the slightest changes in the former can catalyze changes in the latter" (Jacobson-Tepfer, 2020, p. 77). When societies had built their shields but started to interfere intensively with processes on a geological timescale, they did so based on a modern idea of time, which no longer included those early and immediate experiences from generations ago. Generally speaking, modernity's dominant singular temporality is primarily characterized by a dominant and linear vision of newness. Even though concepts, such as justice or belonging, carry temporal components, they are usually bound to rather short timeframes, and counter-temporalities inhabiting ruin or destruction or non-Western concepts of rhythms and endurance could hardly compete, resulting in temporal dislocation (Ginn et al., 2018, pp. 213–214; Serres, 1995). At the same time, the vast ignorance of deep-time interactions, for example, the burning of fossil fuels, led to industrialization and enabled, at least in Western societies, the growth of democracy and economic prosperity. However, staying with the example of fossil fuels, these developments come with "subterranean geologic debt" (Yusoff, 2018a) that conditions freedom in the future. The consequences of climate change limit the range of political options, opportunities, and flexibility available within a given context, namely, the room for maneuver: "deep time is not therefore antecedent to the present but continues to organize and differentiate arrangements of energy and matter . . . to place current concerns into a much larger flow of planetary history and futures, nudging deep Earth forces to disrupt our received narrative strategies and moral imaginaries and in so doing provincialize Anthropocene narratives" (Ginn et al., 2018, pp. 216–217; see also Clark et al., 2016). In general, energy flows, in the context of heat, light, or motion, as they appear in the burning of wood, wind turbines, or solar radiation cycles, are a prime example of how deep-time interactions are based on twofold agency (Walker, 2021). Therefore, deep-time agency, regardless of being human or nonhuman, can "take anywhere from seconds to eons to do their work" (Clark, 2011, p. 201).

Third, why are deep-time encounters politically relevant? As previously explained, they are not only co-made by humans and influence public as well as private life, they also shape political room for maneuver. This is barely recognized, as political infrastructures are adjusted to anthropogenic timeframes without taking deep-time interactions into account. While the recognition of processes taking place within geological timescales is a challenge for the human experience, it is not a risk to politics as such. The risk is that the massive scale is misinterpreted as rendering politics obsolete; however, it rather forces a rethink of government (Bonneuil & Fressoz, 2013/2016). Remembering that politics is based on the plurality of people (Arendt, 1950/1993, p. 9), the opposite is true; deep time and the multiplicity of cultural origins as the necessity and basis of democratic politics are not only compatible, they are necessarily interconnected and condition one another (Zielinski, 2013, p. 7). What has been perceived as the "outside" or the "far away" does not exist anymore; it is, at least implicitly, already part of politics (Chakrabarty, 2018, p. 29; Westermann & Rohr, 2015). Instead of including children or grandchildren in political decision-making, novel ways to establish trusteeship for democratic processes, combined with forms of political kinship with radical alterity of nonhuman temporalities, need to be investigated and implemented (Haraway, 2016; Thompson, 2010). As I will discuss later, the recognition and inclusion of deep-time interactions is about to become the next democratic revolution (see Section 3.2).

The challenge is not only to characterize deep-time interactions as a new basis on which to think and do politics, but also to focus the existing political infrastructure on anthropocentric conceptions of time (Bjornerud, 2018, pp. 11–12). This includes fiscal years as well as parliamentary terms that award short-term acting – ranging from less than a year or a congress period to only one human life span – with the prime example that "[b]y discounting at standard rates, the inevitable collapse of the living systems on this planet several hundred years from now could be counterbalanced by relatively trivial economic gains in the immediate future" (Randall, 1988, pp. 219–220). Thus, politicizing biodiversity in a deep-time manner has to acknowledge that it "is perhaps the most precious planetary resource, for which the timescale of replenishment, known from past mass extinctions, is tens of millions of years" (Langmuir & Broecker, 2012, p. 580). Destroying biodiversity translates into distinct forms of temporal violence as it is an act of irreversibility (Hamilton et al., 2015). However, the processes taking place within geological timescales without human influence might diminish political room for maneuver, for example, if an incident of the magnitude of the Great Oxygenation Event 2.45 billion years ago were to occur. Visible in rust belts (iron oxides) around the Earth, this event brought along the extinction of many species. Deep-time interactions are thus woven into our everyday life, and they should equally be woven into planetary politics. With the conceptualization of the politics of deep time in this Element, I aim to take a first step toward the formulation and institutionalization of such politics of deep time.

2.3 Normativity in Deep Time

If deep time is the realm in which societies interact with processes taking place on a geological or even cosmic timescale, questions arise as to how societies should act in and shape this realm, which shaping is desirable and which is undesirable. In a procedural manner, these questions require a permanent and democratic process of societal self-understanding, as discussed later (see Section 3.2). My aim here is to identify core issues that such a process needs to deal with. As of now, "[t]alk of ethics renders banal a transition that belongs to *deep time*, one that is literally earth-shattering. In deep time, there are no ethics" (Schmidt et al., 2016, p. 2). I, however, suggest there are, or should be, and I try to understand which entry points might pave the way to these core issues in three steps.

First, no deep-time interaction is neutral (Yusoff, 2018b). Even though, in terms of planetary transformations initiated by humans, one might say that "We are as gods and might as well get good at it," at times, it might be better to withhold our powers, as we first "HAVE to get good at it" (Brand, 1968, p. 2; 2009, p. 1). Different geological formations are foundational to enable the extractive economies of the Anthropocene, even colonialism and slavery are closely tied to geologic realities. The dispossession of land, a common theme in world history, was never executed without coercion. The American slave trade was from its beginning closely entangled with mining activities, such as the Brazil Gold Rush, with half a million slaves forced to mine (Machado & Figueirôa, 2022; Yusoff, 2018b, p. 14).

No deep-time interaction is neutral and the way this is circumscribed influences the way one thinks about and acts in the world. What "a social geology" can thus help to understand is "not to 'humanize' geology so much as it is to understand how the languages that already reside within it are mobilized as relations of power – and how a different economy of description might give rise to a more exacting understanding of geologic materiality that is less deadly" (Yusoff, 2018b, p. 12). Who or what is valued as having agency matters. The agency of those who were forced to bring the Anthropocene into being, in addition to the agency of those who forced them and the agency of geological realities themselves, needs to be explicated. Then, it becomes clearer who is responsible for which kind of deep-time interactions, making the normativity of deep-time interactions visible, including the close ties of geological material and bodily work, and alternative futures (Yusoff, 2018b, p. 103). The decision, for example, on whether a geological event like the Anthropocene began in 1950, which represents an understanding of history in line with a "Western understanding of time according to which the past is resolutely over and done with rather than being simultaneously here in the present" (Smail, 2021, p. xi), or whether one takes into account the deep roots of the Anthropocene that may be traced back up to ca. one million years ago when fire was first used as a tool, determines whose agency is included and whose is not, whose sufferings are overheard and whose benefits accumulate (Ellis et al., 2016).

It is, therefore, hardly surprising that an analysis of the power relations of deep-time interactions enables to see a new kind of violence that is often extremely slow but can also rupture in brief moments of time. Most closely related to such an understanding of violence is the notion of slow violence:

> By slow violence I mean a violence that occurs gradually and out of sight, a violence of delayed destruction that is dispersed across time and space, an attritional violence that is typically not viewed as violence at all. . . . Climate change, the thawing cryosphere, toxic drift, biomagnification, deforestation,

> the radioactive aftermaths of wars, acidifying oceans, and a host of other slowly unfolding environmental catastrophes present formidable representational obstacles that can hinder our efforts to mobilize and act decisively.
> (Nixon, 2011, p. 2)

Such slow violence hits poor people the hardest and devaluates their ability to become involved in the decision-making of deep-time interactions, such as the sixty-seven "nuclear tests" from 1948 to 1958 on the Marshall Islands and the consequent radiological and chemical violence, which serves as a prime example (Cohen, 2018; Nixon, 2011, pp. 6–7). Deep-time agency was taken from the people of the Marshall Islands, making their homes uninhabitable. Slow violence causalities are a major challenge to scientific, legal, political, and representational practices, as they extend toward the invisible, contrasting the temporalities of planetary biophysics with the human's senses and respective brain's neural circuitry rather than being wired to the human's life span (Nixon, 2011, pp. 8–13). However, in terms of deep time, an extension from slow violence toward what might be called "deep-time violence" needs to be developed. An extension from slow violence to deep-time violence would involve examining the often-hidden effects of human interference on processes within geological and cosmic timescales. Deep-time violence would encompass the recognition that human activities can disrupt and damage the intricate web of life and the complex processes that unfold over extended periods of time, affecting not only future human generations, including a possible successor of *Homo sapiens*, but also the nonhuman world. This concept recognizes that harm and violence can occur over vast periods of time, often hidden or overlooked because of the limitations of our immediate temporal perspectives. An illustrative example of deep-time violence is that of volcanoes. Today, there is growing evidence that anthropogenic climate change also affects volcanic activity (Doocy et al., 2013; Fasullo et al., 2017; Kutterolf et al., 2013; Sigmundsson et al., 2010). Evidence from the past helps to predict what might happen in the future due to human activity. The rapid retreat of ice sheets that preceded the Holocene increased seismicity by way of a postglacial expansion of the lithosphere: the Earth's crust became less solid, and magma would find its way to the planet's surface more easily. In Iceland, the frequency of volcanic eruptions increased by more than 10 times by the end of the last glacial period, and volcanoes from such bygone geological eras can also be found in Antarctica (Palmer, 2020; Wilson et al., 2012). In a normative sense, the question arises which concepts can provide guidance to include the agency of those excluded from decision-making on deep-time interactions and transform deep-time violence into deep-time opportunities through deep-time politics.

Second, the core normative purpose of the politics of deep time is to understand how deep-time interactions define our world, in order to formulate search spaces for who or what, and how to include them in the making of these worlds and establish what should be possible within them. There is no ideal way to handle deep-time interactions which stretch across unhuman dimensions (Chakrabarty, 2019, p. 25). To open the search space, the distinction between the geological and the biological at least becomes fluid. The distinction made since the nineteenth century fostered great progress, but at the same time, this must be understood as having been grown historically and rooted in the Western world view: the distinction between what is perceived as cosmos, geos and bios is culturally and geographically connotated. Many, particularly Indigenous, means of understanding across the world work without it (Bobbette & Donovan, 2019; Whatmore, 2006). What once was alive becomes inorganic, what once was inanimate can be brought to life. To an even greater extent, matter, in this vein, can be defined as frozen action, that is, "not snapshots of preexisting things frozen in time – caught in the act as it were – but rather condensations of multiple material practices across space and time" (Barad, 2007, p. 360). Humans are thus no longer the pride of creation, even though they might assume responsibilities hardly attributable to the nonhuman. For this reason, the multispecies and even the more-than-human perspectives with their relational ontologies are particularly insightful for the politics of deep time (Celermajer et al., 2021; Srinivasan & Kasturirangan, 2016).

All this requires political experimentation and even artistic speculation, with novel arrangements that take deep-time interactions seriously (Bakke, 2017). When it comes to the question of whom to include in the shaping of deep-time interactions and thus the creation of new worlds, the demos needs to be defined in the broadest sense: it is necessary, for example, to experiment with ways of including, probably sensor-based and with the support of machine learning, the processes taking place within cosmic timescales in the political process or, more anthropocentrically, ways of representing different geological eras in the political arena, for example, through spokespersons. In a similar manner, the Indigenous Maori requested that the Whanganui River, Mount Taranaki, and the Te Urewera Forest be granted personhood rights in New Zealand. What these spokespersons bring to the fore are forces of reproduction such as earth care labor and they emphasize the "agency of reproductive and subsistence workers as those subjects that, through both daily practices and organized political action, take care of the biophysical conditions for human reproduction, thus keeping the world alive" (Barca, 2020, p. iii). Life in this sense results "from human-independent evolutionary processes stretching back into deep time" (Palmer et al., 2014, p. 429), putting the idea of cosmovivialism at center stage: "[C]osmovivir may be a proposal for a partially connected commons achieved without canceling out the uncommonalities among worlds because the latter are the condition of possibility of the former: a commons across worlds whose interest in common is uncommon to each other" (De La Cadena, 2015, pp. 285–286). Extracting a chronopolitan perspective from this, the world is "an evolving system of changing temporalities. It presupposes the global present, but transcends it by opening up to alternative pasts and futures, and also to the diversity of intersecting rhythms of life" (Cwerner, 2000, p. 337). Necessarily, and particularly in terms of societal deep-time interactions with processes taking place within a cosmic timescale, the chronopolitan perspective has in the sense of cosmovivir been spatially perceived as multiplanetary to ensure, for example, that the resources of the solar system, ranging from solar energy to minerals possibly extracted from asteroid mining, are available to everyone on Earth in potential deep-time interactions (Losch, 2019).

Third, regarding both of the aforementioned substantive issues – no deep-time interaction is neutral and we need to understand who or what should be included in their shaping – each results in a distinct task: to shape deep-time interactions that keep both humans and nonhumans safe and alive (realization of habitability) and to establish a political process that guides the way in which deep-time interactions contribute to worldmaking (the realization of democracy in a deep-time manner).

The latter is, from a deep-time perspective, best addressed by enabling "natality" of the political again and again anew (Arendt, 1958). This should result in a constant refining of democracy as a government of, by and for the people and the planet (drawing on Lincoln, 1863) to realize positive freedom as the possibility to unfold one's life and realize one's purposes (Berlin, 1969), as I explain later in more detail (see Section 3.2).

The first-mentioned aspect, the realization of habitability, is, on the other hand, concerned with the fundamental protection of negative freedom, which entails the absence of barriers, constraints and obstacles (Berlin, 1969). The fact that Earth is habitable is not just due to its appropriate distance from the Sun. Other requirements for enabling and sustaining life include the presence of sufficient energy to sustain the metabolism and reproduction of living beings and the availability of water and carbon to build complex molecules (Cockell et al., 2016). To further enable habitability in a deep-time manner, responsibilities declare themselves due to the interferences humans created with the conditions of habitability. This encompasses responsibility for (in)animate areas, such as those zones least transformed by humans, but also for artificial objects created by humans, such as nuclear waste and hybrid objects characterized by flows between humans and their environment, such as infrastructures.

However, while planet Earth possesses the characteristics of habitability, it is indifferent to life at the same time. According to Hegel's reflections, when living through the so-called year without a summer in 1816, it is thus "an assertion that we should not – under any circumstance – let the contingent rumblings of the Earth undermine hard-earned political rights and entitlements" (Clark & Szerszynski, 2021, p. 122). In a situation where the potential undermining of political rights and the disregard of responsibilities arising from human interferences with planetary habitability can never be ruled out, the politics of deep time has, in a normative sense, also been thought of as a security project (Mathews, 2020, p. 74). At a very fundamental level, this combines three different features: dealing with potential collapse, avoidance of existential risk and deep adaptation.

Currently, systemic disaster discourse predominantly builds on an idea of sustainability, which still advocates prevention, yet is not ready to factor in either an ecological or a social collapse in a deep-time manner. Dealing with the real possibility of collapse, where basic needs may no longer be met, requires forms of anticipation that are scientifically grounded, socially and existentially enriched, and ethically oriented to understand how to behave if "the state of emergency threatens to become the normal state" (Beck, 1992, p. 79; see also Servigne et al., 2021; Servigne & Stevens, 2020).

To avoid a situation in which the capabilities to deal with collapse are needed, the prevention of existential risks becomes essential. This especially includes events that combine high impact with low probability: these are often characterized by irreversibility and range from full-scale nuclear war to sun storms and asteroid impacts, to the release of engineered pathogens and the complex risks of artificial intelligence (Bostrom, 2013; Ord, 2020). Of course, there is a need to democratize existential risks and to avoid dangerous long-termism that plays off current against possible future generations (Crary, 2023; Cremer & Kemp, 2021; Torres, 2021).

In the case of such existential risks or the interplay of core risks, including "failure of climate change mitigation and adaptation; extreme weather events; major biodiversity loss and ecosystem collapse; food crises; and water crises" (Future Earth, 2020, p. 15) leading to great instability, considerable adaption is inevitable. This includes preparation for "starvation, destruction, migration, disease and war" based on "resilience, relinquishment, restoration and reconciliation" (Bendell, 2018, pp. 12, 21). Thus, it seems as if the realization of negative freedom in a deep-time manner would have to go hand in hand with a strong democratic counterpart to also enable positive freedom.

3 The How: Politicization of Deep Time

> Addressing these sorts of "deep-time" questions forces us earth system governance scholars to broaden our empirical and methodological portfolio and move beyond conventional analyses of environmental regimes, institutional interactions, and global governance studies, which tend to focus on much shorter timeframes.
>
> (Galaz, 2019, pp. 115–116)

In this section I analyze *how* deep time is currently politicized. It starts by explaining the scattered landscape of the mostly implicit politics of deep time, discusses the democratic challenges of deep-time interactions, and, by drawing on the definition of deep time (see Section 2.1), develops an analytical framework to map and analyze eras of the politics of deep time.

3.1 Scattered Politics of Deep Time

The politics of deep time is neither coherent nor fragmented; it is scattered as it is vastly inexistent, which is why conventional conceptual, methodological, and empirical tools within Earth system governance research reach their limits (Galaz, 2019, pp. 111–116). More precisely, I agree with the rationale that current societies are approaching a new stage of complexity. Throughout history, humans have created organizations for their survival in order to face an increasing number of new challenges. At the beginning, organizations focused on basic human needs like food and shelter at local level, then on settlements and their interactions at regional level. During the industrial revolution national coordination became necessary, which was followed by the rise of international organizations to enable the exchange between nation-states at global level. The last few decades saw the rise of a largely network-based world society, which corresponds with the rise of transnational organizations (Battersby, 2017; Shinohara, 2016). At this point, it seems that the end of spatial differentiation has been reached, as political institutions, even if partially fragmented, exist at and across all levels.

My rationale in terms of deep time is as follows: I argue that societies, after having differentiated for centuries within the dimension of space across multilevel political architectures, now enter the dimension of time, which requires novel multitemporal political mechanisms and institutions (Hanusch & Biermann, 2020, p. 33). The temporal scope of multitemporal politics ranges from ultrafast algorithmic processing in microseconds to the eras of geological and cosmic processes (Galaz, 2019, p. 122).

While the number and intensity of deep-time interactions have grown almost exponentially over the last few decades, the respective political architectures have not formed, and responsibilities have not been assumed. This is hardly surprising as the exponential growth of deep-time interactions started overwhelmingly at the same time as the Great Acceleration in the 1950s (Steffen et al., 2015) which could be observed in the simultaneous rise of various socioeconomic indicators (e.g., world population, transport, telecommunication) and Earth system measurands of human activity (e.g., domesticated land, ocean acidification, loss of tropical forests). At this point, however, the constitutions of most states contributing most to these developments, such as the Constitution of the United States from 1788, as well as those of international organizations, such as the Charter of the United Nations from 1945, were already written, without being aware of the need to govern deep-time interactions. Changes to such documents, which would need to include deep time, for example, taking nonhuman agency into account (see Section 2.2), would be fundamental and would require a new constitutional moment. The fact that such a constitutional moment will be fostered top-down by political leaders or demanded bottom-up by grassroot movements is, at the moment, rather unlikely, given the fact that deep time is barely an issue of political contestation which reveals a limited deep-time literacy throughout world societies. The politics of deep time is from a current point of view more likely to emerge organically, with specific organizations as prototypes, such as the Svalbard Global Seed Vault, as an example of an existing organization that is meant to last essentially forever to preserve genetic information of cultural crops, or final depositories for nuclear waste in the making, that might in sum and over timescale up internationally (Ialenti, 2020; Westengen et al., 2013). The challenge for any politics of deep time then becomes the required longevity of respective organizations (Hanusch & Biermann, 2020). Not only are political arrangements needed that are capable of processing deep-time interactions into politics, while accounting for various temporalities such as speed, rhythm, and acceleration, but deep-time political arrangements also need to last over very long periods of time. As of now, organizations that have survived – only – a few millennia or centuries are hard to spot, and some of those that exist, such as the Japanese monarchy or the Catholic church, can only partially serve as a model for the democratic societies of the twenty-first century and beyond. Thus, what is needed is a focus on the politics of deep time enabling continued democratic connections between people and planet in a deep-time manner.

3.2 Deep Time as a Challenge for Democracy

Deep-time interactions are political and the sheer number of possibilities of how they can be arranged, on the one side, and their importance for life on planet Earth, on the other side, demand adequately equipped democratic politics. Before I outline how the democratic politics of deep time might look,

I explain why deep time is a substantial challenge for the current democracies' presentism and the kind of requirements for the democratic politics of deep time can be deduced from this. I also discuss why approaches that aim to include future generations fall short when it comes to the huge timespans of geological or even cosmic processes.

First, why is deep time a challenge for democracy? While some challenges for democracies through deep time are rather explicit and easily observable, others are more implicit. The explicit challenges can be traced back to the human lifespan, the respective incumbent interests, and, accordingly, the timeframes inscribed in societal institutions: they lead to short-termism and a presentist bias in democratic politics, as, for example, in the democracy-climate nexus (Hanusch, 2018). Most obviously, this includes election cycles of four or five years and even more autocratic states align their political planning to similar timeframes, including China's five-year plans. Decade-long democratic planning, as in plan approval procedures when major infrastructures are developed and implemented, for example, the installation of renewable energies and respective power lines, is only a fraction of the time in contrast to deep time. Obviously, there is no quick fix, as adjusting election cycles to the timeframes of geological processes, which confront "the political with forces and events that have the capacity to undo the political" (Clark, 2014, pp. 27–28), would only give rise to some sort of deep-time technocracy.

Even more difficult, yet providing a clear set of tasks, are the implicit deep-time challenges for democracy. In the study of time, a distinction is made between future presents and present futures (Adam & Groves, 2007; Koselleck, 1979/2004; Luhmann, 1976). Present futures are an anticipation of the future from the present point of view and allow for planning for various expected changes. In contrast, future presents – as the actual futures once enacted – are largely subject to the unknown, as they put unexpected change at center stage. Therefore, future presents are those that might be of primary importance when it comes to deep-time interactions and timescales so vast that uncertainty clouds the imagination as to what the interests of any future beings might be. They simply can longer be accessed using the techniques and methods of imagination, forecasting, or anticipation. In other words, two fundamentally different modes of futuring can be distinguished: futuring that addresses expected change or futuring that addresses unexpected change (Monda, 2018; Szántó, 2018). Within deep time, both are of relevance but future presents, in a deep-time manner, are central to the question of how the politics of deep time can be democratized: how can unexpected change taking place within geological timescales be incorporated into democratic politics?

Second, why do approaches that focus on the inclusion of future generations not tackle deep-time challenges? In short, even though experimentation relating to the inclusion of future generations in democracies' institutional design might be promising for the generation of children and grandchildren (Boston, 2016; Köhler, 2017; Smith, 2021), this approach cannot cope with future presents, as these are characterized by unexpected change and are thus largely unknowable to human imagination (Whiteside, 2018). This includes attempts to reform legislatures, for example, through proxy representation of future generations, by including the veto votes of sub-majorities against legislation that impacts future generations negatively or by introducing offices of future generations, such as in Israel, Hungary, or Wales, which, as elite institutions, often lack democratic legitimacy in the first place. This also includes mini-publics in the form of citizens' assemblies and citizens' juries, such as those practiced in the French La Convention Citoyenne pour le Climat or the Scottish Climate Assembly, which aim to realize high degrees of deliberation, independence, and diversity through random selection. While the latter approach might, to a certain degree, enable forms of deep-time thinking through deliberation, random selection itself is only one way of representation that has no direct connection to deep-time interactions. Moreover, future generations might identify other selection procedures that may be more democratic and better equipped to deal with deep-time interactions. In addition, the idea of random selection is not neutral but is rooted in an Athenian tradition of doing democracy. Both, deliberation and random selection are based in present futures, as they hardly allow future communities to self-select and self-represent with their own selection strategies from their time. This makes random selection a practice that potentially leads to exclusionary standards, demanded under unequal conditions, as future generations cannot be present in the here and now (Morán & Ross, 2021); the politics of future generations in general could be an excuse strategy to act in the here and now (Humphreys, 2022).

Third, what are possible pathways to deal with deep-time interactions democratically? In short, I propose a combination of deep-time literacy, inclusion of the nonhuman and a trustee conception of sovereignty.

Starting with *deep-time literacy*, it is necessary to empower people to exercise their citizenship in a deep-time manner. Deep-time literacy can be characterized as an expanded awareness and understanding of the interactions, effects, and implications of human interference with processes taking place within geological or cosmic timescales. This includes the cultivation of a mindset and knowledge base that goes beyond the immediate present of the current and a few generations in the past or future and considers the temporal interwovenness of the long-term influences and consequences of human's actions, building on such approaches as technological citizenship or futures literacy (Frankenfeld, 1992; Miller, 2018). The core democratic approach is to enable literacy, in order that knowledge of deep time becomes part of thinking and acting in the daily lifeworld and in democratic politics; in this vein, people become educated in deep time and can act accordingly as democratic citizens. Such deep-time literacy of the long ago past is particularly palpable in fossils, strata, or meteor craters, which help to make this distant past tangible to the human senses. A deep-time literate person obtains a sense of "timefulness," defined as the ability to locate ourselves within eras and eons, rather than weeks and months, comprehending "an acute consciousness of how the world is made by – indeed, made of – time" (Bjornerud, 2018, pp. 5–6; see also Wood, 2018). This also includes re-learning the languages of lost worlds, such as Native American myths which are embedded in the landscapes: in contemporary Europe, such land-connectedness is mostly lost or turned into fairytales and bizarre spirituality, with deep-time amnesia prevailing in its highly industrialized and digitized societies (Du Cann, 2021). All this can contribute to an actively enacted, deep-time culture and respective bottom-up norm formation, which is to date barely existent, but is tentatively approaching in the arts (Saltmarshe & Pembroke, 2019). This could bring about a more temporally aware society, and ultimately a poly-temporal world view (Toulmin & Goodfield, 1982).

Democratizing the *inclusion of the nonhuman* requires, for example, the already tested yet still anthropocentric construction of human proxy representations or, in a more experimental manner, the direct and technology-based inclusion of processes taking place within geological timescales, with which current societies are interfering (see Section 4.2). A prime example is the interaction between ice sheet cycles and the deposition of nuclear waste, the final depositories of which need to last for ca. one million years. During the 100,000-year ice age cycles, ice sheets thousands of meters thick are formed, the sea level fluctuates by 120 m and considerable land subsidence and uplift occurs due to the ice load, all of which can severely deform and damage a repository and therefore jeopardize its safety (Ganopolski & Brovkin, 2017; Willeit et al., 2019). More generally, a proxy representation of ice sheet cycles, macroevolution, seasonality, and similar processes means to monitor and represent processes taking place within a geological or cosmic timescale in an evidence-based manner, through nonhuman processes that have similar transformative potential as human societies in regard to planetary change. If we look back at the history of democracy, a central factor that democratized democracies was the process of inclusion. When those, who were thought to lack agency and who were seen primarily as a human resource for labor or reproduction, such as slaves, non-whites, or women, were included in democratic decision-making, democracies became more democratic, new possibilities opened up, and the quality of life improved. Which possibilities would open up if we now included planetary processes taking place within geological timescales in democratic decision-making? What, at first glance, seems to be a thought experiment has already become a real democratic trial run for the inclusion of the nonhuman, as previously explained, in terms of the Indigenous Māori speaking about a river, a forest or a mountain. Similar pathways of representation for other nonhuman entities in a deep-time manner, such as ice sheet cycles or the process of evolution, are thus equally feasible. Moreover, novel, sensor-based technologies together with machine learning algorithms are imaginable as a tool for more direct forms of inclusion. Together, they could possibly generate a certain kind of "will" of planetary forces proceeding within geological or even cosmic timescales (Bakker, 2022; Bridle, 2022). However, these experiments can face serious practical challenges, as the proposed solutions might not work as intended, might lead to secondary non-intended effects, or might not fit into the context for which they were intended; therefore, they may not be as forceful as expected. This can be partially observed in Ecuador and Bolivia, where the adoption and practice of buen vivir and The Law of the Rights of Mother Earth are an ongoing experiment and learning process.

Lastly, the *trustee conception of sovereignty* allows for every future generation to live with great degrees of autonomy. A trustee conception is fundamentally different from conceptions of future generation representatives that only allow for present futures imagined by the current generations. A generation living in 1,000 or 200,000 years from now cannot be imagined, and the kind of interactions they will practice with processes taking place on a geological timescale, which might range from continued large-scale geoengineering to interplanetary space mining, cannot be anticipated. A trustee conception of sovereignty is also different from what might be termed deep-time justice, as it requires no reference to individuals, other cultures, or future societies. Thus, the main idea of a trustee conception of sovereignty is not focused on the content of deep-time interactions but on the process: "Present sovereigns can represent future sovereigns by acting as trustees of the democratic process" based on the general principle that "present sovereigns should act to protect popular sovereignty itself over time" (Thompson, 2005, p. 248). Accordingly, representatives are neither trustees of the interests of future citizens nor of a specific form of democracy, but trustees of the conditions empowering future citizens to establish a democratic process and make collective choices in a democratic manner (Thompson, 2005, p. 249). A trustee conception of sovereignty would come with a set of institutions. These include, for example, posterity impact statements of governments on the potential effects of future sovereigns on democratic capacities or constitutional conventions to align these with the changing values of a democratic process (Thompson, 2005, pp. 256–259). In this vein, the challenge of rendezvous in deep-time interactions is addressed, namely, to coordinate and synchronize societies that act or come to exist at different points in time, in order to have a chance to repeatedly readjust their respective politics of deep time (Saraç-Lesavre, 2021). The fundamental idea that even "[t]he making of any constitution is potentially an act of inter-temporal tyranny" (Thompson, 2005, p. 251) is nothing new, and, as a consequence, the democratic process itself needs to be renewed until eternity.

Amidst the tumult of the French Revolution in 1789 and shortly before his return to the United States from Paris, where he acted as ambassador, Thomas Jefferson expressed his thoughts on "[t]he question whether one generation of men has a right to bind another" in a letter to James Madison (Jefferson, 1789/1958). At this time, the Western world was busy negotiating the nature of modern democracy against the backdrop of a new conception of time: the space of experience was increasingly separated from the horizon of expectation so that the future seemed open, the present no longer God-given, and the past as one of many versions (Koselleck, 1979/2004). This was taken as a carte blanche, along with a misconstrued idea of progress, but Jefferson insisted "that the earth belongs in usufruct to the living" (1789/1958). It cannot be utilized at will but must be handed over to the next generation in a state at least as good as we received it, and without debts. The future president derived a radical proposal from these deliberations: the constitution and laws were to expire with those who willed them into being; otherwise, the dead would govern the living.

Behind this idea is the assumption of contingent futures. However, since the unborn cannot be represented politically, the independence of generations has to be ensured. Each of these must be free to begin anew, "that is the actualization of the human condition of natality" (Arendt, 1950/1993, p. 178, own translation). Jefferson's insights point to the temporality of democracy. Each generation must obtain self-efficacy through an act of foundation and negotiate whether path dependencies, like the subsidization of fossil fuels, are compatible with the principle of usufruct. By letting go of the old and creating their own constitution, new generations would reinforce the democratic way of life in a deep-time manner. Jefferson closed his letter to Madison with the remark that his proposal will "[a]t first blush ... may be rallied, as a theoretical speculation: but examination will prove it to be solid and salutary" (1789/1958). Indeed, it is high time that we engage in such an examination. In planetary times, the belief that our democracy will forever be stable is just plain naïve, much more so than the ongoing experiments that aim to renew it with confidence and courage, such as experimentation with democratic forms of the politics of deep time.

3.3 Analytical Scheme of the Politics of Deep Time

Based on the rationale substantiated in Sections 2 and 3, the aim of Section 4 is to investigate the deep-time interactions, in order to provide an outline for the main traits of future politics of deep time in Section 5. The focus is thus on identifying the main characteristics of deep-time interactions, to understand the kind of politics of deep time that are needed. The respective guiding question for the analyses of empirical cases is: what characterizes the deep-time interaction and the respective politics of deep time?

To answer the guiding question, I first provide a definition of politics and categories of investigation, second, an explanation of the methods used and third, a justification of the case selection.

First, the categories of investigation result from the definition of politics applied. Contemporary definitions of politics often have a strong focus on a certain issue, most often on the role of the state and the government. In contrast, the largely unmapped territory of deep-time interactions and the respective exploratory character of this Element, based on an inductive research design, best align with a broad and open definition of what is understood as politics in the politics of deep time.

A starting point for such a broad definition is best found when considering the etymological root of the word politics, which comes from the ancient Greek term, "polis," referring to the city-state. Politics in that sense encompassed all aspects and matters related to the city-state. The term "politics" was thus an integrative term that did not differentiate, for example, between the government and the state, the state and society, society and personal life, or any of the aforementioned and morality. This old understanding of politics is surprisingly close to the most recent attempts in political science that aim to capture the vast array of activities influencing politics, such as the term governance. This is to say that the term "politics" was, in previous decades, largely reduced to the activities of the state and government, as the dominant view was that these are the only or most important categories to be investigated. This does not and never did hold true in a world where not only non-state actors but also nonhuman forces influence societal, political, and planetary change.

This study aligns with and aims to reinvent the original understanding of politics. Accordingly, politics in this study is defined as all aspects and matters related to deep-time interactions. The analytical categories I apply to investigate all these aspects and matters related to deep-time interactions are: dynamic, agency, and architecture. Indicators of this exploratory and indicative study are defined openly in order to be able to depict a variety of characteristic values. In other words: the aim is to map the core elements of the politics of deep time, a largely unmapped territory.

The category dynamic explicates the constantly changing nature of deep-time interactions that needs to be governed. Here, I aim to investigate the core characteristics of deep-time interactions, understanding them as "earthly multitudes" or, in other words, as the various connections which diverse societies establish with the "planetary multiplicity" of an ever-changing planet (Clark & Szerszynski, 2021, pp. 171–172). Therewith, I focus on the bidirectionality of deep-time interactions, meaning I take into account the societal influence on processes taking place within geological and cosmic timescales, as well as the influence these processes have on human societies. Respective indicators include: processes within geological and cosmic timescales relevant to societies; opportunities and constraints for societal (re)actions; feedback loops and iterative processes of the interaction; potential tipping points or critical junctures through (inter)planetary events; mechanisms that recognize, politicize, and address interactions.

The following two categories, agency and architecture, are also conceptualized in an analytical manner through indicators. Yet, as explained in the previous sections, an intentional politics of deep time does not currently exist. For this reason, in the analyses of the case studies, the aim cannot be to investigate the current status of the politics of deep time. Instead, the aim is to explicate in an anticipatory manner the way in which these should be practiced under the normative principles of democracy and habitability, as part of intentionally set up political arrangements. Adequate political arrangements would have to enable the autonomy of all possible future generations to exist on a habitable planet.

The category, agency, investigates which human and nonhuman entities actively influence deep-time interactions: on the one side, human societies have agency to shape processes taking place within geological and cosmic timescales and on the other side, these processes themselves exert nonhuman agency influencing human societies. This requires novel definitions in terms of "the complete *integration* of human and nonhuman agency [that] breaks down conceptual barriers between humans and their 'surroundings' and integrates them in a complex understanding where agency is diffuse, interactions are dynamic, and boundaries become blurred" (Biermann, 2021, p. 64). Respective indicators include: the potential roles of human and nonhuman entities in the politics of deep time; the symbolic significance of human and nonhuman entities; the attribution and recognition of more-than-human agencies from mental models to legal frameworks to actual participation; the power relationships between human and nonhuman entities; the co-constitution of political outcomes by human and nonhuman entities.

The category, architecture, can be defined as the design, structure, and physical manifestation of political systems, institutions, and spaces. Architecture is seen as co-constitutive with the agencies that enact within and through it, while human and nonhuman entities constantly and simultaneously (re)create the architectures. In the context of deep-time interactions, political architectures are a core category as they potentially shape the course of civilizations over very long time periods. The respective indicators include: formal and informal values, beliefs, attitudes, rules, and norms; the distribution of power and resources in decision-making processes; the degree of inclusivity and representation; the role of information, knowledge, and communication; the capacity and ability to adapt to changing circumstances.

Second, in terms of methods, the qualitative meta-analysis, which integrates the primary and secondary sources of deep-time interactions, is exploratory and inductive, in order to identify characteristic elements for the formulation of a basic conceptual framework of the politics of deep time. The diversity of cases investigated in this study may lay the foundation for broader and possibly an increasing number of standardized quantitative studies.

The guiding question – what characterizes the deep-time interaction and the respective politics of deep time – is therefore divisible into four sub-questions: (i) Why is this case relevant for the politics of deep time? (ii) What are the core dynamics, that is, the aspects and matters related to deep-time interactions, that need to be governed? (iii) Who should – based on the principles of democracy and habitability – possess agency to change the interaction? iv) How should the political architecture – based on the principles of democracy and habitability – be developed?

Third, not only are the planetary challenges facing contemporary societies complex, they also interfere with one another forming a "polycrisis" (Tooze, 2022). The interconnectedness and interdependence of risks have the potential to trigger a chain reaction and lead to a systemic crisis. As described previously, climate change, the occurrence of severe weather, significant biodiversity loss and ecosystem collapse, food shortages, and water scarcity must, therefore, be understood not only each as a risk for itself, but their interaction and interdependence create the greatest challenge (Future Earth, 2020).

To understand and address this complexity their fundamental material, spatial, and temporal dimensions have to be investigated. The latter renders them a genuine case for the study of the politics of deep time. Thus, the cases investigated in the following section are complex and cannot be explained by a deep-time perspective alone, yet they are genuinely characterized as interactions between contemporary societies and processes taking place within a geological or even cosmic timescale. Each of the deep-time interactions takes place in another of planet Earth's interrelated spatial spheres or even outer space and represents a material process occurring within geological or cosmic timescales. Again, the aim is to inductively gather diverse empirical insights and recognize patterns that will allow me to identify a broad range of elements to compile a rich conceptualization of the politics of deep time. This results in the following case selection scheme (see Table 1).

Based on the criteria of diversity regarding the spatial sphere and the material process, the cases selected are as follows: the influence of biogeochemical soil formation on the election results in Alabama is presented as an example of deep-time interactions with the geosphere; the way in which karst aquifers enable societal flourishing showcases the impact of the hydrosphere and its freshwater cycles on societies; the interrelationship between the cryosphere, its glacial cycles, and human activities is analyzed by considering the search for a final depository for nuclear waste in Germany; the Pleistocene Park in Siberia, Russia, is presented as an example of human interaction with the biosphere and the process of evolution; solar cycles interfering with the Earth's magnetosphere are a case of possible space weather mitigation; the potential terraforming of Mars as planet (re)formation calls for respective multiplanetary politics or a Martian government.

Of course, most of the deep-time interactions investigated exist in the past, present and future. For example, biogeochemical soil formation or solar cycles took place before the existence of *Homo sapiens*, are currently taking place, and will also take place in the future. Deep interactions thus make new causal connections between widely separated periods in the Earth's history.

Table 1 Case selection of the politics of deep time

spatial sphere	material process	deep-time interaction
geosphere	soil formation	plankton vote democrats
hydrosphere	fresh water cycle	living with karst aquifers
cryosphere	glacial cycles	nuclear rendezvous
biosphere	evolution	rewilding the Pleistocene
magnetosphere	solar cycles	space weather mitigation
outer space	planet (re)formation	terraforming Mars

4 The What: Explicating the Politics of Deep Time

> In deep time, everything is provisional. Bones become rock. Sands become mountains. Oceans become cities.
>
> (Gordon, 2021, p. 10)

This section focuses on *which* concrete cases are in need of deep-time politics. Diverse empirical insights and patterns are identified that provide the basis for a draft conceptualization of the politics of deep time in the Section 5. Within each deep-time interaction, I first investigate the case in terms of its overall political relevance; second, I explicate the core characteristics of the dynamics of the deep-time interaction; and third, I outline the required political arrangements in terms of agency and architecture. The cases may sound unfamiliar or even irrelevant at first, as a deep-time perspective and the cosmic timescales are novel in the realm of political research. Yet, as I will demonstrate, these are of utmost significance for numerous people across many generations currently as well as in the future.

4.1 Plankton Vote Democrats

First, the Black Belt region of Alabama is an area of political confrontation. The region is named for its dark, fertile soil and is made up of predominantly African American communities. The area has a rich history of black activism and struggles for civil rights, including the Montgomery bus boycott and the Selma to Montgomery march. The large African American population, historically disenfranchised and marginalized in the political process, has faced voter suppression and gerrymandering, aimed at diluting the voting rights of African Americans. In recent years, efforts have been made to mobilize the Black Belt electorate and increase the political participation of African Americans, as they have the potential to be a significant voting group in statewide elections. In addition, the Black Belt region is often cited as an example of racial and economic inequality in the United States. The region is one of the poorest in the country, with high rates of poverty and unemployment, and limited access to healthcare and education. The Black Belt thus serves as a reminder of the systemic inequalities that persist in the United States and the need for policies to address them. As I will show, the evolution of the Black Belt is associated with the politics of deep time as the Cretaceous evolution can be related to contemporary electoral behavior, a case of deep interaction between societies and soil formation processes within the geosphere.

Second, what are the core dynamics of deep-time interaction? The case in focus is a deep-time interaction between biogeochemical soil formation processes and local societies, which manifested in a region now known as the US state of Alabama. More precisely, the interaction takes place in the so-called Black Belt that stretches through Alabama toward northeastern Mississippi. It is a core agricultural region in the United States, where cotton was the main crop until the American Civil War. Later diversification with corn, soybeans, and beef cattle, among others, took place.

The deep-time interaction can be clearly visualized by comparing different maps of the region over time. To start with, the reason for the fertility of the soil dates back to the Late Cretaceous, lasting from 100.5 to 66 million years ago, when the eastern part of North America formed a continent called Appalachia (see Figure 8).

Figure 8 North America in the Late Cretaceous Period, around 75 million years ago.

Source: Ron Blakey © 2013 Colorado Plateau Geosystems Inc.

On the southern coast of Appalachia, sediments were deposited. The ocean temperature and the sea level were higher back then; plankton died near the coast, resulting in massive chalk deposits from their skeletons. As time went by, oceans cooled down, receded and the chalk, now above sea level, made the soil in the former coastal region alkaline, fertile, and dark. The so-called Black Belt was formed. Today, this former coastline runs through Alabama and continues east and west into neighboring states. Older rocks can be found north of the Black Belt, as this part of Alabama was already part of the landmass before the Cretaceous. The landmasses of southern Alabama, however, were formed after the Black Belt and are, therefore, dominated by younger rocks (Dutch, 2020; McClain, 2012).

In the 1800s, white farmers set up cotton plantations along the Black Belt (Silkenat, 2022). As many of the Indigenous peoples, who initially lived there, were killed, the farmers forced the enslaved people of African descent to till the fields and pick cotton. The vast exchange of flora, fauna, and culture across the Atlantic after 1492, known as the Columbian Exchange, included the violent enslavement and displacement of 11.7 million Africans: European colonists preferred enslaving Africans over the Indigenous population, as Africans were assumed to be better suited for hard labor, as they showed a greater immunity to diseases brought to the Americas by European colonizers (Crosby, 1972; Mann, 2011). In this context, the agency of biogeochemical soil processes comes to the forefront. Counterfactually argued, without dead plankton on the southern coast of Appalachia, leading to the fertile soils of the Black Belt, slaves would not have been needed to plant and pick cotton, at least not in this region.

At the beginning of the twentieth century, ca. six million African Americans moved northward as part of the Great Migration, also known as the Black Migration. However, many also stayed, and as a result, the majority of people in counties in the Black Belt today are African Americans. They stayed, not only because they were too poor to leave, but also because they had worked the soil for generations and now saw an opportunity to own this land after the enslavement was over, at least legally (English, 2020). The soil that once was the reason for their exploitation was believed to become a source of prosperity for them, even though this did not come true.

The close connection between soil and society is evident from the fact that the name "Black Belt," once used to describe the fertile soil of the region was later used to refer to the large number of Black (former) slaves living there:

> I have often been asked to define the term "Black Belt." So far as I can learn, the term was first used to designate a part of the country which was distinguished by the colour of the soil. The part of the country possessing this thick,

dark, and naturally rich soil was, of course, the part of the South where the slaves were most profitable, and consequently they were taken there in the largest numbers. Later, and especially since the war, the term seems to be used wholly in a political sense – that is, to designate the counties where the black people outnumber the white. (Washington, 1901, p. 3)

"Black Belt" is thus a very vivid linguistic expression of how societies are entangled with biogeochemical processes forming soil.

As African Americans were prevented from voting, even a century after the Civil War ended, the Black Belt with its African American majority became the center of the Civil Rights Movement. One major historical event, the Selma to Montgomery Marches in 1962, took place on what is called the Selma Chalk, a geologic formation the name of which dates back to the Cretaceous. Voter suppression of African Americans is still in existence today, as the Supreme Court suspended the Voting Rights Act in 2013. Nonetheless, since most African Americans voted and still vote for Democrats, whereas the remainder of Alabama is dominated by Republicans, the outline of the Black Belt is still clearly visible in the maps of election results. In other words and illustrated in Figure 9, dead plankton from 100 million years ago favored the ruthless exploitation of enslaved people, which after centuries of oppression led to the Civil Rights Movement and as a consequence, to voting rights for African Americans, resulting in contemporary political maps which reflect biogeochemical processes from the distant past (Dutch, 2020; McClain, 2012): "The long-conceded regional identity of the Black Belt roots no more deeply in its physical fundament of rolling prairie soil than in its cultural, social, and economic individuality" (Gibson, 1941, p. 1).

Similar patterns of deep-time interactions can be found in different places, with the German Ruhr region serving as an example. As a major region for coal mining, the distribution of coal across different areas correlates with contemporary voting patterns. Coal reserves in the Ruhr are not evenly distributed but become deeper the further north one goes. Consequently, the industrialization of the Ruhr region proceeded in a northerly direction, which is referred to as "northward migration." Today, primarily the southern regions, where the wealthier sections of the population live, have been transformed into renaturalized areas, while the northern regions are home to the remains of the former coal mines and the communities that depended upon them until a few decades ago. As a result, there is a remarkable discrepancy in voting behavior, with the majority of the northern population voting Social Democrat by comparison with the south.

Third, what political arrangements are needed to enable deep-time habitability and autonomy? The case of the Black Belt demonstrates the interwovenness of today's societies with biogeochemical processes: these occurred millions of years ago but reveal themselves currently as deep-time interactions. Agency, which is relevant for conceptualizing the politics of deep time, is neither primarily embodied in the biogeochemical process causing fertile soil nor the societies that live on the site where the process took place. Instead, it is the relationship between both societies and biogeochemical soil processes within geological timescales, which exerts agency as a deep-time interaction. Regarding the potential politics of deep time, the question then arises, how this interaction needs to be governed to enable many to thrive. However, as a respective political architecture was missing for decades, if not hundreds of years, deep-time interactions were not steered with the public interest in mind; instead, agency was only exerted by a few predominantly white men who misused it for their own purposes. Obviously, such deep-time interactions, although intuitively believed to be in the past, also require politics in the here and now.

Political architectures are needed to prevent the potential that accumulated over geological timescales, such as fertile soils in the Black Belt, from getting exploited and, in addition, from being used to exploit humans. Instead, such potential, in accordance with the usufruct principle, would probably need to be governed as deep time and thus planetary commons. In addition, deep-time reparations, similar to war reparations, which would compensate the degradation of fertile soils by exploiting humans for one's own benefit, are a tool to, at least partially, compensate for undemocratic worldmaking and, even more importantly, to allow for hopeful futures (Táíwò, 2022). Such deep-time reparations could, in the case of the Black Belt, include the restoration of overused agricultural areas – even though the effects of past agricultural land use may be irreversible in historical timescales (Dupouey et al., 2002) – the compensation of communities, whose ancestors were enslaved and exploited or the remodeling of ownership structures. In this sense, the first colonial agricultural use of the Black Belt is not very different from the practices in the first agrarian states in Europe: the accumulation of domestications, including plants, captives, and women in patriarchal families served to gain control over reproduction (Scott, 2017). Furthermore, the remodeling of ownership structures includes the abolition of disadvantages resulting from property laws, which particularly affect low-income rural populations, mostly African Americans and Native Americans who often die without a will, in highly fractionalized ownership (legally known as "tenancy in common"), contributing to perpetual poverty in the Black Belt (Dyer et al., 2008). Racial discrimination is a persistent feature of contemporary societies, particularly from a deep-time perspective,

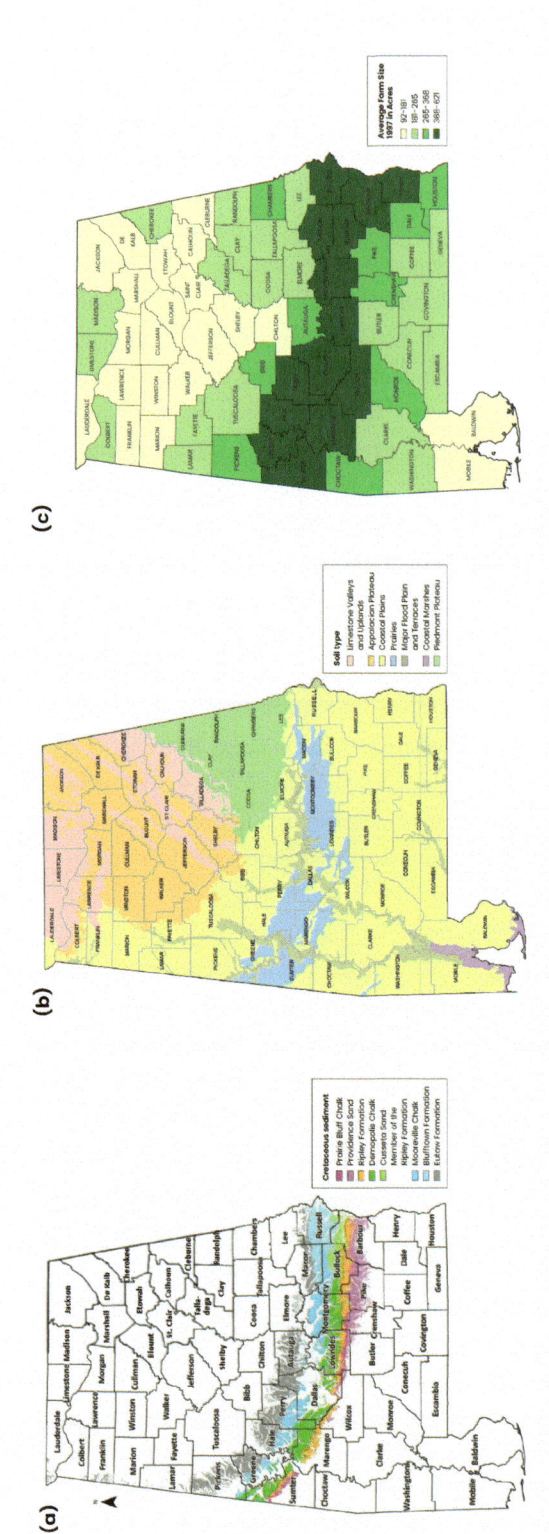

Figure 9 Processes on geological timescales shape human societies: exemplarily the influence of sediment distribution on elections is shown for the US state of Alabama. (a) Shows the distribution of cretaceous sediments across the state. (b) Depicts the distribution of soil types in Alabama. Rich prairie soils are distributed along the lines of the sediments, earning the region the name "Black Belt." Fertile soils favor the establishments of farms, therefore, as a consequence (c) the average farm size is largest within the belt. In (d) and (e) a similar pattern can be seen in the dispersion of the slave population in 1860 and the African American or Black population in 2010. Finally, (f) illustrates the results of the 2020 presidential election. The Black Belt is the only region in Alabama that voted for the presidency of Joe Biden. Please note that the 2020 election constitutes the ideal case and other election results do not replicate this pattern as clearly. However, a general trend is observable in most elections.

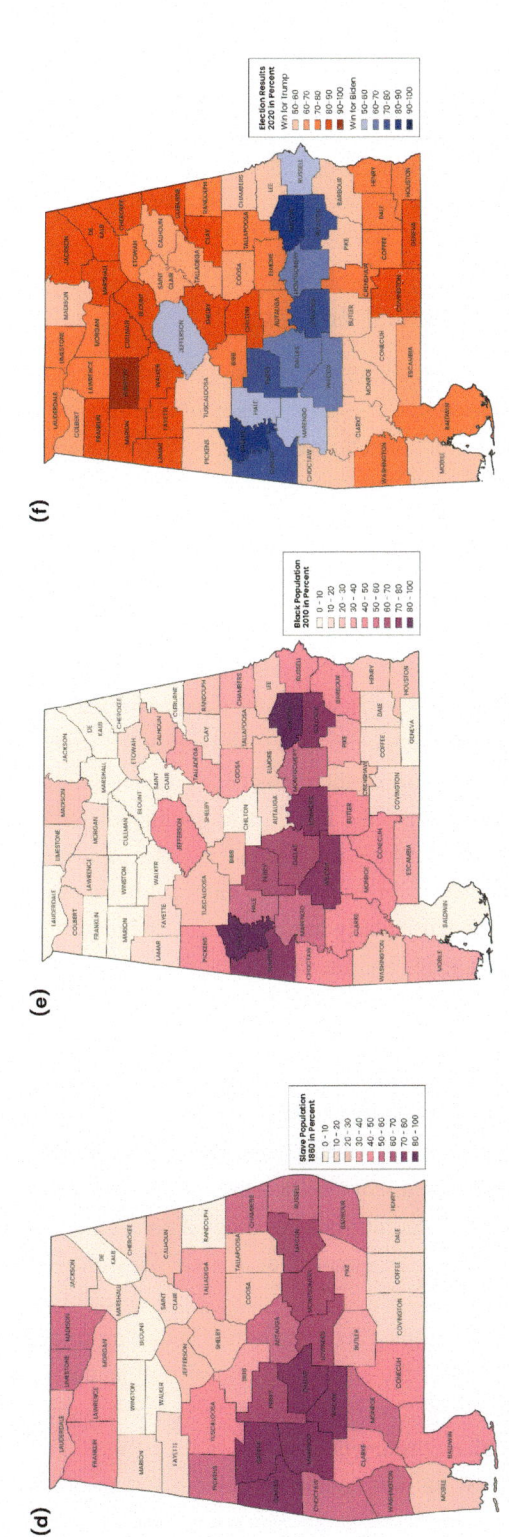

Figure 9 Source: (a) Image used with the permission of the Alabama Museum of Natural History, published in Ikejiri et al. (2013); (b) image, and (c) data: courtesy of Alabama Maps– aproject of the University of Alabama; (d, e) data from the US Census Bureau; (f) data from the Alabama Secretary of State; county and state boundaries: (1) for 1860 (d): georeferenced map reproduction courtesy of the Norman B. Leventhal Map & Education Center at the Boston Public Library; (2) for present day (c, e, f): Database of Global Administrative Boundaries (GADM).

[t]he planet did not begin with these [racial] divisions, and there is no reason why they should persist as a taxonomic bedrock, a rationale for carving up the world's populations into discrete units. There is such a thing as a preracial planet. Its reference point is geological time, at the tail end of which Homo sapiens emerged, a small tawdry band, its survival uncertain, standing or falling as a species, and only as a species. (Dimock, 2008, p. 177)

4.2 Living with Karst Aquifers

First, the Kendeng Mountains in Indonesia constitute a significant ecological and cultural area that has been affected by large-scale cement production and is thus characterized by political confrontation. One of the main problems associated with this industry is the extraction of limestone from karst aquifers, which has been shown to cause severe environmental damage, including the depletion of water resources and the pollution of groundwater. Political disputes around this issue involve a range of stakeholders, including local communities, environmental activists, government officials, and cement companies. At the heart of politics surrounding karst aquifers in Kendeng is the tension between economic development and environmental protection. Cement production has been touted as a major driver of economic growth and job creation in the region, which has been a key reason for government support of the industry. However, this has been at the expense of ecological integrity, the health, and livelihoods of the local population. For this reason, activists and local communities have opposed cement production and the extraction of limestone from karst aquifers. From a deep-time perspective, this case is particularly important, as it demonstrates a deep-time interaction between societies and the freshwater cycle within the hydrosphere.

Second, what are the core dynamics of the deep-time interaction? Before I analyze the concrete case, I will explain the general functioning of karst aquifers. They are, as of now, quite unfamiliar to political science research, not only because they are invisible below the surface and are hardly accessible and measurable even for karst researchers, but also because they were formed within geological timescales that only become visible through a deep-time lens and might otherwise have been taken for granted as a stable background condition.

Karst landscapes form unique and complex ecosystems due to the "processes of karst dissolution, the permeability of the solutionally developed landscape surface, the presence of a well-developed and open subsurface, fewer surface streams, and an overall calciumrich environment" (Stokes et al., 2010, p. 377; see also Hartmann et al. 2014). Karst aquifers contain 13 percent of the world's groundwater and supply ca. 25 percent of the world's drinking water, constituting almost the sole groundwater source for large cities such as Vienna, Rome, or Damascus (Onac & van Beynen, 2020). They exist around the world and form large underground systems that cross national borders (Stevanović, 2019; see Figure 10).

Karst aquifers form in soluble bedrocks where the chemical processes of water containing CO_2 create underground fractures and channels. Once the water exits the underground system, it forms springs with a high degree of hydraulic volatility (see Figure 11). Karst aquifers have formed over many geological eras, for example, the aquifers in the Blue Grass region in Kentucky date back to the Ordovician age, 430 to 500 million years ago, which is why the respective caves are also archives of our past life. The widely ramified underground system makes them highly vulnerable, not only to chemical and microbiological pollutants, which can easily enter and manipulate the aquifer system, but also to adverse interactions with human infrastructure, such as dams, mines, tunnels, or railways, due to problems with sinkholes, flooding, or leakage (De Waele et al., 2011, p. 1; White et al., 2018; Zapletalová et al., 2016).

Karst aquifers are intriguing sites, demonstrating how humans interact with the results of processes that have evolved over millions of years. For example, the Leang Tedongnge cave, a karst cave in Sulawesi, Indonesia, contains the oldest art work known to man, a 45,500-year-old Sulawesi warty pig (Brumm et al., 2021; Uomini, 2016). As karst aquifers exist worldwide, karst aquifers and human culture have been interacting since the beginning of human history: For example, in Australia, karst aquifers have been used for partially still unknown cultural practices from at least 10,000 years ago (May & Tacon, 2014; Spate & Baker, 2018), in Belize, the Maya altered karst aquifer geomorphology for their agriculture which is likely to have contributed to their downfall (Beach et al., 2015), and in Greece, karst aquifers are even identified as the "hydrogeological basis of civilization" (Crouch, 1993, pp. 63–82).

A key characteristic of deep-time interactions with karst aquifers, which can be of particular importance for the politics of deep time in general, is the development of "geomyths" and "hydromyths": These contain insights into natural history, as myths often concern humanized geo- and hydrological events and therefore include relevant information to help understand how processes taking place within geological timescales influence contemporary societies (Clendenon, 2009a; Mayor, 2005; Montgomery, 2012; Murphy, 2023; Vitaliano, 1973). Myths are not fairy tales or fables; "they are the science of cultures that do not verify the truth by means of experimentation" (García & Gaviro, 2017, p. 186; see also Chakrabarti, 2020). To identify real events within myths, heuristic techniques, namely, euhemerism, can be applied. These

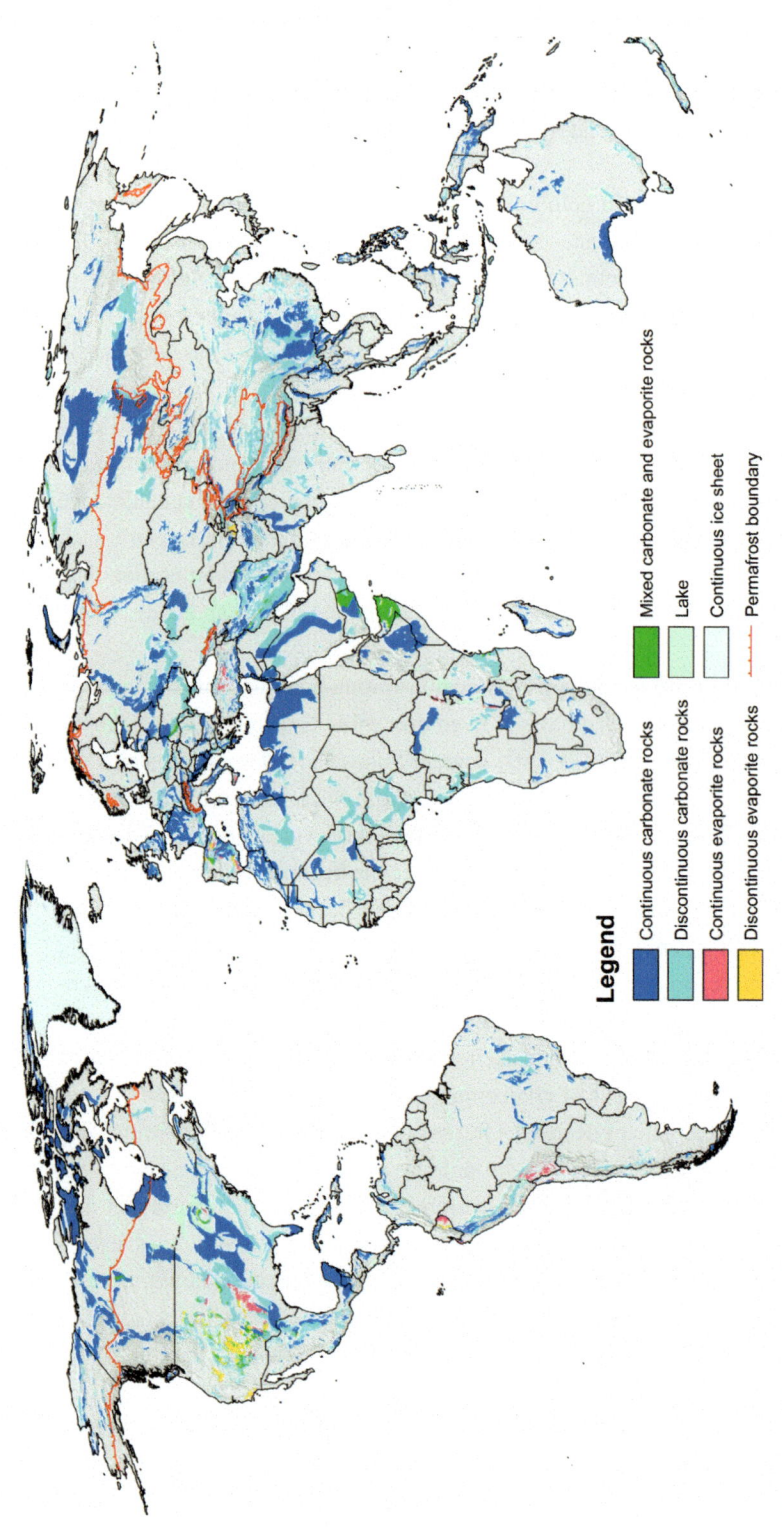

Figure 10 The Global Karst Aquifer Map shows the distribution of the different types of karstifiable rocks which represent potential karst aquifers.

Source: World Karst Aquifer Map (WOKAM) © Bundesanstalt für Geowissenschaften und Rohstoffe (BGR), International Association of Hydrogeologists (IAH), Karlsruhe Institute of Technology (KIT) and UNESCO (2017); map modified by Goldscheider et al. (2020, p. 1665), CC BY 4.0

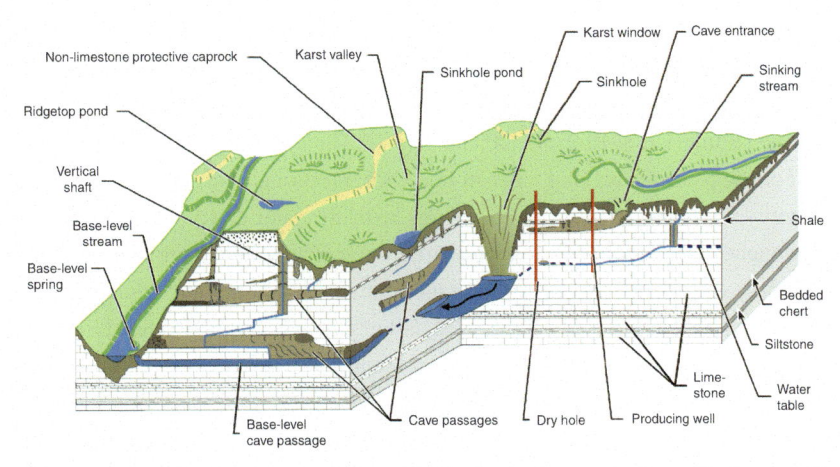

Figure 11 Typical physiographic and hydrologic features of a well-developed karst terrane, using the example of the Western Pennroyal Karst in Kentucky, USA.

Source: Currens (2001) © Kentucky Geological Survey, University of Kentucky

techniques allow to situate dates, places, and persons of myths in a multidisciplinary manner in actual historical settings: "In a world full of hazards, myths confirm that a pattern exists" (Back, 1981, p. 257).

Myths, in particular, in karst regions deal with the appearance and disappearance of creatures in caves, sinkholes, or springs. Ancient civilizations and Indigenous people narrating these hydromyths largely perceive "water not only as a resource, a place, a flow or a particular belief, but it integrates everything into a single worldview, which is what we can call water culture of that place" (García & Gaviro, 2017, p. 189). Especially in Greece and Turkey, research has shown that the extrapolations of mythological explanations, for example, regarding local karst hydrogeology and the karst water transportation systems over dozens of kilometers, withstand modern scientific experiments (Clendenon, 2009b, 2009c).

The deep-time interaction in the Kendeng Mountains is also shaped by such myths that are narrated by the Indigenous people living in North Central Java, Indonesia, who call themselves Sedulur Sikep, "the friendly ones," yet are sometimes referred to as Samin people (Benda & Castles, 1969; Korver, 1976; Maliki, 2019; Putri, 2017). The Sedulur Sikep came to prominence due to the activism of Samin Surosentiko, born in 1859, who opposed Dutch colonialism and preached resistance and free access to the forest as a common resource as traditionally practiced by Indigenous communities (Benda & Castles, 1969, pp. 234–235). The Sedulur Sikep movement practices the fusion of some other cultural accounts, particularly a Western perspective, opposing qualities, such as spirituality and materialism, with their ritual practices extending toward the *longue durée,* aiming for practices even beyond generational thinking and acting (Sumarlan & Rumpia, 2021).

For the peasants leading a life in line with traditional Indigenous order, their identity is rooted in close relationships with the land and in respective self-sufficiency. The myths of the region are mostly oral and thus provide a holistic perspective rather than precise measurements, as they can be found in some Greek mythologies. Karst aquifers are part of a larger integrated worldview that centers around balance and harmony, referring to Mother Earth as the central figure. Thus karst aquifers, as well as the Kendeng Mountains area to which they belong, are both a medium expressing spirituality and a freely accessible common resource. The landscape of the Kendeng Mountains is characterized by teak trees and rice fields, crossed by rivers and numerous springs and caves. The Kendeng Mountains are, in the perception of the Sedulur Sikep, more than a physical landscape. Accordingly,

> [s]ome sacred sites used by the Sedulur Sikep to have meditation-like rituals or prayers to God for specific purposes, as well as the water springs and caves that irrigate Sikep houses and rice fields receive special blessings and offerings. This way of treating the elements of the Kendeng Mountains, especially those that directly benefit the people, is hoped to help maintain and conserve their sustainability and express the people's gratitude to Mother Earth. (Putri, 2017, p. 310)

Like Mother Earth, the Kendeng Mountains do not know administrative borders in view of the Sedulur Sikep, but "stretch across five regencies as one single entity" (Putri, 2017, p. 319).

Knowledge of the karst aquifers and their importance for water catchment became threatened when cement companies, namely, Indocement, a subsidiary of the German HeidelbergCement (HC), started extracting limestone from the Kendeng Mountains to produce cement. As a result, not only are huge amounts of energy needed in the sintering process, but also the karst aquifers were, are, and might be irreversibly damaged by destroying the structure of underground systems and, therefore, changing water paths, which leads to erosion and results in flooding and landslides. Policies advocating living in a deep-time manner with karst are lacking; rather, it can be recognized as "problems surrounding the legal politics of limestone exploitation in the karst area, which often ignore the legal politics of the safety of the womb and the life chain of the karst ecosystem, which results in inequality between generations of the karst area" (Konradus, 2021, p. 2). The Sedulur Sikep oppose the exploitation of the

Kendeng Mountains in the form of the "Jaringan Masyarakat Peduli Pegunungan Kendeng" (JMPPK), the "Network of the people who care about the Kendeng Mountains," particularly because they interfere with its water resources and thus endanger agricultural activities. In 2021, one of the activists named Gunarti wrote an open letter from the Kendeng Mountains stating:

> 11 years ago, life at the Kendeng Mountains was still undisturbed. In the morning the roosters loudly welcomed the new day. The birds were chirping. And from many springs of the mountain water flowed in streams into the valley. We lived in peace and balance. ... The year 2021 also began for us with floods. To this day, we have flooded rice fields in the Kendeng Mountains, where the harvest has failed. It is a disaster that is man-made. The flooding of villages and fields happens because people continue to rape Mother Earth. They do violence to her by mining and cutting down forests until whole swaths of land are left bare. And then there is nothing left to stop the water. If there are no more fields to grow anything, what will you eat? If there are no more farmers like us, who will provide you with food? ... You should honor and preserve the earth, which has given you everything you enjoy in your lives. You should not make the earth angry. (Gunarti, 2021)

Third, what political arrangements are needed to enable habitability and democracy in a deep-time manner? Three aspects come into focus: to better understand karst aquifers and their agency, to develop architectures that treat the irreversibility of their karst destruction, and, finally, to understand the past and formulate novel narrations that can be passed on to numerous generations.

To start with, the hydrogeological process that forms the karst aquifer has agency in its own manner. Yet, this process and most karst aquifers resulting from it around the world are barely understood in terms of their functionality, particularly the way in which humans interact with it. Not only are karst aquifers barely visible as they mostly lie underground and are physically hard to explore, but they are also taken for granted: karst aquifers and many other natural resources have been exploited long before humans even began to understand their importance for the long-term functioning of the Earth's systems. The world's first karst aquifer map was completed just a few years ago (Chen et al., 2017; see Figure 10), a karst disturbance index was developed once but has not been continuously and comparatively applied (Van Beynen & Townsend, 2005), and, to the best of my knowledge, only one genuine institute devoted to karst research and outreach, the National Cave and Karst Research Institute in the USA, exists worldwide. To raise awareness, the International Union of Speleologists even declared 2021–2022 the "Year of Karsts and Caves," outlining: "Karst aquifers are the most complex, least understood, most difficult to model, and most easy to contaminate water supplies. They are often able to rapidly transmit pathogens and chemicals tens of kilometers undetected to vital human and ecological water sources. ... Many governments do not recognize caves and karst at all or fail to recognize their importance" (International Union of Speleology, 2021). Thus, to develop the evidence-based politics of deep time, there is a need for more knowledge on the agency of the hydrogeological process that forms karst aquifers which, in turn, deliver freshwater for hundreds of millions of people. This includes political and social sciences that have just recently been identified as relevant for the future of functioning karst aquifers (Younos et al., 2018, p. xi).

Regarding the irreversibility of karst aquifer destruction, the respective political architecture is needed to ensure their protection. First and foremost, a non-extraction agreement is required that applies to all those deep-time interactions which are characterized by missing regenerative capacities for essential hydrogeological processes: such processes which are of upmost importance for the habitability of the planet and the autonomy of future generations. Even a non-alteration agreement of karst might be considered, as karsts are, despite their fragility, discussed as potential locations for storage units of carbon capture and storage technologies. A respective karst law, building upon circumstances as in Indonesia, would have to understand the hydrogeological process, which formed karst aquifers and their respective landscapes, as an irreplaceable and complete whole that enables life: they are characterized by many springs, enable the formation of habitats and the development of cultural heritage, and provide the basis of economic prosperity (Konradus, 2021, p. 5). In addition, karst aquifers and the forests, which often form around them, constitute a common for the Indigenous people that needs to be protected. In this regard, positive inspiration might be drawn from the Charter of the Forest of 1217 – the counterpart of the Magna Charter – which survived almost 800 years, establishing the right for everyone to access and use the common as their needs require (Linebaugh, 2008; Robinson, 2020). Yet, even in practical terms, the sustainable use of karst aquifers for the needs of twenty-first-century industrialized societies seems to be possible. In Miskolc, Hungary, for example, deep thermal karst water is used for the operation of the largest geothermal heating plant in Central Europe, with a capacity of 60 MWt (Miklós et al., 2020).

Lastly, one feature of the politics of deep time that becomes particularly important when treating extremely long time periods are narrations and especially myths, based on evidence, which can provide a stable framework for the long-lasting politics of deep time. Myths allow to uncover and retell processes taking place within geological timescales, as they extend the persistence of information beyond the human lifespan, outlasting numerous generations into the future or, in ultimo, even getting passed down through all human lifetimes. Myths can be the narrative circumscription of complex relationships in deep time. They are needed as cultural techniques to enable treatment beyond the existence of political institutions, as an alternative way of long-term information processing (van der Leeuw, 2020). A starting point for this could be the status of a World Heritage Site by UNESCO. In 2018, 37 karst areas were recognized as World Heritage Sites, with three of them having a mixed, natural, and cultural status, namely, the Tasmanian Wilderness in Australia, the Hierapolis-Pamukkale in Turkey, and the Pyrénées-Mount Perdu in France and Spain (Trofimova, 2018). The evidence that past myths transported to the presence is numerous, such as myths by Native Americans about a fissure under Seattle having led to a major earthquake 1,100 years ago; a myth by the Thai Moken with regard to tsunamis carrying a warning that when the water moves back, a major wave follows – in fact, the tsunami of 2004 caused the death of 300,000 people but almost none of the Moken died – or a myth about the Nyos Lake in Cameroon warning of the "deadly breath" of the lake, which, in the form of sulfur dioxide and carbon dioxide emanating from an extinct volcano below the lake, killed 1,700 people yet no Indigenous people in 1986. However, such myths do not only prevent death, they also relate springs to well-being, such as the sacred wells which can be found from Madagascar to Indonesia to Siberia (Ray, 2019). To create new myths, one can also draw on insights from archeology as a deep-time laboratory which demonstrates, for example, a Classic Mayan cosmocentric worldview, where not only people and animals, but also plants, rivers, stones, or clouds contribute to the maintenance of their world (Guedes et al., 2016; Hambrecht et al., 2020; Lucero & Gonzalez Cruz, 2020). However, myths need to be evidence based; otherwise, they might form, for example, stories about a young Earth, a belief held so strongly that it is even institutionalized in the Institute for Creation Research in Dallas or the Creation Museum in Petersburg in the United States. Thus, there are also potential drawbacks of applying deep-time myths as part of the politics of deep time. Myths can be subjective and open to interpretation, leading to different understandings, which undermine their effectiveness across cultures and potentially marginalize the perspectives of minorities. They also carry the potential of being exploited or manipulated for individual interests in the here and now. As with almost all potential tools relating to the intentional politics of deep time in the making, deep-time myths must also be seen as an experiment that has to be handled with care, in order to identify evidence-based symbols, metaphors, and archetypes in a democratic manner.

4.3 Nuclear Rendezvous

First, the search for a final repository for nuclear waste in Germany is a highly political issue per se, as Germany has a strong green movement, which fought a decade-long fight against nuclear power. The process has been ongoing for decades, and various sites have been proposed and then discarded due to local opposition and environmental concerns. The latest plan is to store the waste in a deep geological repository, built deep underground to prevent possible radiation leaks. However, finding a suitable site has proven difficult, as many communities oppose such a repository. As a result, the search for a suitable site for the repository has become a source of political tension and a Federal Office for the Safety of Nuclear Waste Management was established in 2014. The decision to find a final repository for nuclear waste is thus not only a technical decision, but also a deeply political one. Particularly from the perspective of the politics of deep time, this involves a profound interaction between societies and glacial cycles in the cryosphere.

Second, what are the core dynamics of the deep-time interaction between glacial cycles and energy carriers, in particular, uranium, exemplified by the German case of seeking a final depository for its nuclear waste? In general, energy carriers, human development, and Earth system dynamics are closely entangled. Coal and oil, for example, enabled industrialization, put forth unprecedented wealth increase, provoked labor movements, degraded landscapes, and facilitated climate change. In turn, uranium enabled large-scale electrification, triggered green movements, brought about nuclear disasters, and required final depositories for one million years. Thus, the energy carriers that human societies choose not only influence the deep-time interaction with glacial-interglacial cycles, especially when choosing to burn fossil fuels, but the glacial-interglacial cycles also influence the way human societies must handle energy carriers, as in the case of nuclear waste.

The Quaternary, which started ca. 2.6 million years ago, is the current period of the Cenozoic Era. Within this time period, the climate alternated between two extremes, ice ages and warm periods, with glacial-interglacial cycles spanning ca. 100,000 years. During the last ice age, which ended ca. 10,000 years ago, the global average temperature was five to six degrees and the sea level was 130 m lower than today, due to 3-kilometer-thick ice sheets. In total, ca. 32 percent of the Earth's surface was covered with ice, compared to around 10 percent today. Throughout the ice ages, the ice masses of the Arctic, the Antarctic, and the mountains advanced greatly within a few hundred years, covering large parts of Europe, Asia and North America. Traces of the ice ages include, for example, moraines, glacial scars, and erratic boulders. Furthermore, ice sheets play an active role in the Earth's carbon cycle, due to adapted microbial communities, geochemical weathering, and the storage of organic carbon (Wadham et al., 2019). Ice cores allowing for these environmental insights, thus, carry an agency that transforms human's temporal consciousness, situating it in cycles of hundreds of thousands of years (Antonello & Carey, 2017).

One key reason for glacial cycles are changes in the Earth's orbital geometry, resulting from gravitational forces in the Sun-Earth-Moon system. Basically, the shape of the Earth's elliptical orbit around the Sun (eccentricity) changes over periods of around 100,000 years, and as a consequence, this changes the distribution of solar energy on Earth, with ice ages occurring when summer solar irradiation in high northern latitudes becomes minimal. Other contributing forces include, for example, the oceans which act as conveyor belts for heat, such as in the Gulf Stream. Besides these changes within cycles of 100,000 years, cases of rapid temperature drops by 10 °C in 40 years also exist. Such sudden drops in temperature can probably be ascribed to solar activity and were the cause of the so-called Little Ice Age in the seventeenth and eighteenth century. Why is this relevant for the politics of deep time in relation to a final depository for nuclear waste?

Due to the use of ca. 100 million-year-old fossil fuels, human societies are becoming a force with a similar magnitude to orbital variations, creating climate impacts which last for hundreds of millennia (Archer, 2009). Recent models calculating future glacial cycles estimate that the next glacial inception would naturally occur within the next 50,000-90,000 years, but due to the current level of anthropogenic emissions (500 Gt (Pg) of carbon), this is likely to be postponed to at least 120,000 years from today as a result of high emission scenarios (3000 Gt (Pg) of carbon), with estimates suggesting that the next glacial inception will not take place within the next 600,000 years (Talento & Ganopolski, 2021). While the burning of fossil fuels by human societies will impact glaciation in 10 or even hundreds of thousands of years from now, the agency of glacial-interglacial cycles also creates demands for human infrastructure planning in the here and now, which in the case of nuclear waste dictates planning one million years into the future.

The main resource used to generate electricity based on nuclear fission is uranium. The uranium that exists on Earth is likely to have originated from supernovas or the merger of neutron stars ca. 6.6 billion years ago. It, thus, becomes obvious that within deep-time interactions the history of the universe and that of Earth's often intersect several times. Uranium is the main source of heat inside the Earth; it can be found in the Earth's crust and can even be extracted from the seawater of the oceans. Today, the main extraction sites are in Kazakhstan, Canada, Australia, Niger, and Namibia.

While many intriguing deep-time interactions exist in the context of uranium, such as a possible nuclear war leading to an artificial ice age, here, the focus is on the impact of glaciation on nuclear waste disposals (Witze, 2020). To illustrate this deep-time interaction, I analyze the German case.

In accordance with the "Act on the Search and Selection of a Site for a Repository for High-Level Radioactive Waste (Gesetz zur Suche und Auswahl eines Standortes für ein Endlager für hochradioaktive Abfälle (Standortauswahlgesetz – StandAG))," the respective site – in what is today known as Germany – must shield highly radioactive waste from the environment for one million years. Looking back in time, there have been six cold periods with glaciation in Germany in the last one million years (Brosig et al., 2020). In terms of a maximum and based on a glacier in Scandinavia with a thickness of up to more than 4000 m, the thickness of the overlying ice was assumed to be 3,500 m for the North German Lowlands and around 400–450 m for the southern margins. Within the timeframe of one million years and the corresponding 100,000-year ice age cycles, not only did ice sheets thousands of meters thick form, but the sea level also fluctuated by ca. 120 m, and there was considerable land subsidence and uplift due to the ice load, which would have deformed a possible final repository of nuclear waste. In order to evaluate the long-term safety of a repository, it is not only necessary to know the current hydrogeological properties and bounding conditions of a site, including groundwater movement and groundwater temperature fluctuations; their possible changes over the next one million years must also be taken into account.

Based on knowledge from previous ice ages, glaciation was a dynamic process in all areas in Germany affected by the ice age. Material transports of several hundred kilometers were common, and accordingly, their impact on a nuclear waste repository has to be considered. Moreover, the stress change in the crust associated with glacial loading enabled the reactivation of tectonic faults down to a depth of at least 2 km across all areas, up to 300 km from the front of the maximum glaciation, namely, Elster in eastern Germany and Saale in northwestern Germany; therefore, the whole of Germany and thus all possible final depositories of nuclear waste are impacted by glaciation (see Figure 12).

However, much is still unknown. Research projects on modeling the effects of changing external boundary conditions on hydrogeologically relevant parameters are still in progress (Bundesamt für die Sicherheit der nuklearen Versorgung (BASE), 2022b). For example, effects on the hydraulic permeability and porosity of the rock, changes in the pathways of deep water and the distribution of fluids in the subsurface are being considered. Special attention is also paid to the mechanical loading and unloading of the subsurface, due to the cyclical alternation of warm and cold periods.

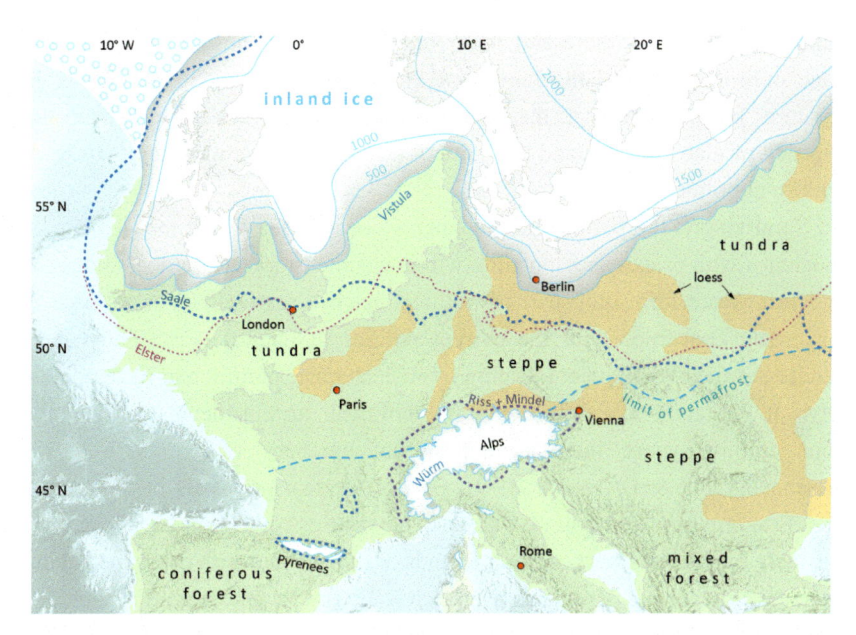

Figure 12 The extent of the ice and its estimated thickness (m) during the three last glacial periods in Central Europe (Elster & Riss glacial maximum: approx. 350,000 years ago, Saale & Mindel glacial maximum: approx. 150,000 years ago; Würm & Vistula glacial maximum: approx. 20,000 years ago). The extent of the land masses shows the dimensions of the European continent with the sea level lowered by 100 m.

Source: Reprinted with the permission of Springer Nature Customer Service Centre GmbH: Springer Cham, The Geology of Germany by Meschede & Warr (2019)

Even though much is still unknown, the present state of knowledge already shows that some fault zones may have been reactivated during the last one million years due to ice loading (Brosig et al., 2020, pp. 45–57). It is assumed that faults and fractures can be reactivated all over Germany, creating additional pathways for deep water to enter the repository. Moreover, model calculations suggest that, regardless of the thickness of the unconsolidated rock cover, further fault zones may be reactivated, at depths considerably lower than 1,000 m. Yet, the case of nuclear waste disposal sites are so intriguing, as they demand that human societies deal with cosmic times in the here and now, putting the temporal uncertainties at the table of contemporary politics in a very concrete manner, as the infrastructures for the repository need to be built currently.

Third, this leads to the question of which political arrangements are appropriate not only to successfully oversee the safe construction of nuclear waste disposal sites in the here and now, but to take into consideration the next one million years and the impact of the respective glaciation cycles. Two features become apparent: the kind of organizations needed to deal with the agency of glacial cycles and the cultural techniques accompanying the site as informal political architectures.

Surprisingly, the political arrangements needed, due to the interference of societies with processes happening within cosmic timescales in the context of nuclear waste form only part of political science research. This research is primarily concerned with the current politics of nuclear waste and barely addresses the overall unique feature of nuclear waste politics and its temporality; it certainly does not explore political arrangements which take one million years into account (Brunnengräber et al., 2015, 2018; Brunnengräber & Di Nucci, 2019). One of the few proposals in this regard refers to organizations dealing with nuclear waste as "deep-time organizations" as these must exist over very long periods of time to deal with processes taking place within cosmic timescales (Hanusch & Biermann, 2020). In Germany, the Federal Office for the Safety of Nuclear Waste Management "performs regulatory, licensing and supervisory tasks for the German government in relation to the disposal, storage, handling and transport of high-level radioactive waste" and is thus the main organization responsible for nuclear waste disposal in Germany (Bundesamt für die Sicherheit der nuklearen Versorgung (BASE), 2022a). We cannot expect such organizations to exist one million years into the future, but a range of principles can be applied in their design to at least allow them to last for as long as possible. These design principles, which historically have been proven to guarantee longevity, include, among others, ensuring the continuous support of elites, no diversification of core activities, and arranging for the public to have ownership and responsibility in decision-making (Hanusch & Biermann, 2020, pp. 31–32). Thus, organizations responsible for nuclear waste should be designed carefully and holistically, not only focusing on the geological and technological aspects of deep-time interactions, but also emphasizing the political and social responsibilities. In other words, making longevity the core and center of every physically existing organization responsible for nuclear waste must be the distinct feature of nuclear waste politics. However, an organization like BASE might cease to exist once its current location is covered by thousands of meters of ice, at the latest during the next glaciation period.

Another main characteristic of nuclear waste politics is, therefore, uncertainty, which demands cultural techniques as a feature of the political architecture, to enable the dissemination of knowledge regarding the final depository. While the politics of uncertainty in a complex world is nothing new, the temporality of the deep-time interaction requires special treatment (Renn, 2008). The three ways to deal with uncertainty in a deep-time manner are rendezvous, cultural techniques, and experiments. Regular and planned rendezvous between different parties at different points in time might help to deal with uncertainty and to adapt the final depository to new circumstances, novel priorities or concerns, necessitating the alignment of temporal conceptions and meeting points (Saraç-Lesavre, 2021). Such temporal arrangements of rendezvous need to be passed on to numerous generations, which rely on cultural techniques lasting beyond the current nation state or even language structures, such as myths, narrations, and practices around the site, similar to those of sacred places (Foley, 2021; Hecht, 2018). This can include symbols placed at the final depository, but also objects dealing with the deep-time interactions visible in the daily lifeworld (see Figure 13 and Figure 14). New monuments may be necessary and may appear, creating semiotic relations that connect the history of humans and the Earth (Szerszynski, 2017; 2020). Nuclear waste politics is thus not only about expertise and physically existing organizations, but the imagination of possible future situations (Ialenti, 2020).

Figure 13 The "Landscape of Thorns" by Michael Brill and Safdar Abidi.
Source: © Sandia National Laboratories, concept by Michael Brill, drawing by Safdar Abidi, originally published in Trauth et al. (1993); available on https://daily.jstor.org/wp-content/uploads/2015/05/NuclearWaste_1050x700.jpg

Figure 14 Nuclear plates ("Atomteller") depict the landscapes surrounding German nuclear power plants. This plate shows the plant Mülhheim-Kärlich near Koblenz. Wall plates in Delft blue are memorials that became a common form of nostalgia in many households in the European regions where windmills existed. Nuclear power plants have taken over from windmills today: energy buildings that shape the landscape. As materialized future memories, nuclear plates forecast and grapple with the nostalgia that is tied to these toxic spaces. They constitute energy buildings which, at least in Germany, are disappearing as did the windmills.

Source: Reprinted with the permission of Andree Weissert and Mia Grau (2019); available at https://atomteller.de/Muelheim-Kaerlich-KMK

Lastly, experiments are needed to treat the uncertainty inherent in deep-time interactions. To prevent the limitation of the character of such experiments, these should be twofold. First, they should address the situation in the here and now. This includes, in particular, the question of international cooperation in the search for final depositories. A brief retrospect clearly demonstrates that it is highly unlikely that the nation states as we know them today will exist within the borders of the current territories in a couple of thousand or even hundreds of thousands of years. A multinational approach, that is, a joint solution for repository projects from several countries, is, to date, only being considered in Latvia, Slovenia, Slovakia, the Netherlands, and the United Arab Emirates, but no concrete plans are currently under way. In Germany, such a solution would even be prohibited by law. Second, experiments should address the distant future, where deep-time interactions persist. A starting point, for example, is the Råången experiment in Sweden (Pelzer et al., 2021). Råången is a piece of land owned by the Lund Cathedral, which became part of a larger urban development plan. The Cathedral aims to plan for a thousand-year-period, not only focusing on financial and land use planning, but striving to engage a reflective and questioning attitude, first and foremost through the arts, to approach the question of how long-term futures can be brought into the present by imagination. Similar approaches need to be experimentally transferred to the case of a final depository.

Even though possible future technological developments might finally enable large-scale fusion power plants to produce a very small quantity of high-level nuclear waste, the nuclear waste that has already been produced by traditional nuclear fission power plants, serves as a reminder of the almost infinite consequences for future generations, when a deep-time interaction is invented without the respective deep-time politics mechanisms being established simultaneously.

4.4 Rewilding the Pleistocene

First, the Pleistocene Park is relevant to the politics of deep time because it represents a proactive approach to changing deep-time interactions with a past epoch, relevant to current and future human societies. The politics around the Pleistocene Park, which is organized as a foundation, can be understood as a collaborative project involving multiple stakeholders, including scientists, policymakers, local communities, and international organizations. The project is guided by a set of principles that prioritize grassland, permafrost restoration and conservation, and climate change mitigation. An important aspect is the use of scientific evidence for decision-making, drawing on ecological research that shows how reintroducing large herbivores can restore grasslands and enhance soil carbon sequestration. Accordingly, ongoing scientific monitoring and analysis to assess the ecological impacts of reintroduced animals and to refine management strategies takes place. The Pleistocene Park has also raised concerns. The reintroduction of large herbivores into grasslands could have unintended, ecological consequences, such as overgrazing and soil compaction, and could potentially impact native species and habitats. There is also uncertainty about the long-term outcomes of the Pleistocene Park. While the project has shown promising results to date, it is unclear whether the restored grasslands will be able to sustain the reintroduced animals for the long-term, or whether the carbon sequestration benefits will be maintained over time. The following deep-time interaction between societies and the process of evolution within the biosphere thus demonstrates how interference with the ongoing process of evolution, which occurs within cosmic timescales, is capable of partially bringing back or, more precisely, introducing bygone species. This is an intentional attempt at manipulating the current and composing novel elements of the biosphere, thus forever altering the evolution of genetic diversity deliberately, not by chance. In this context, the politics of deep time refer to the effects of human interventions on evolutionary processes over long periods of time. This refers to the ways in which humans have affected and continue to affect the biosphere and geosphere, not only through activities such as urbanization, deforestation, and agriculture, but also through the use of technologies such as synthetic biology and Clustered Regularly Interspaced Short Palindromic Repeats (CRISPR).

Second, what are the main dynamics of the case in focus, namely, synthetic biology and its application in the context of the Pleistocene Park? In general, synthetic biology is the biological counterpart of geoengineering. While the latter relates to building artificial (parts of) geospheres, synthetic biology involves designing artificial (parts of) biospheres. Humans have altered the geosphere and the biosphere since their existence, yet to different extents.

Links between interspecific interactions, such as mutualism or competition, and macroevolution, namely, evolutionary change taking place within cosmic timescales, were a largely unintentional and irreversible process until the human species emerged (Hembry & Weber, 2020). Humans were and are profoundly altering the course of evolution, for humans themselves, but also modifying the evolution of others, ranging from the first dogs thousands of years ago to the use of bacteria to produce insulin in (CRISP)-technologies for gene editing (Cohen et al., 1973; Shapiro, 2021).

Almost all land-use changes that can be attributed to humans, from urbanization to deforestation to agriculture, have negative effects on biodiversity (Cordier et al., 2021). Human population size is, with 96 percent accuracy, a predictor for mammal extinction within the last 126,000 years, while other factors, such as (non-anthropogenic) climate change, offer no better prediction than chance (Andermann et al., 2020). In particular, extinctions of megafauna tens of thousands of years ago have diminished species' abundance and overall ecological diversity to a great extent, such as the extinction of the mammoth and the mammoth steppe it co-created (Johnson, 2009). Another, rather recent yet profound, alteration of the biosphere is the so-called "Columbian Exchange," the transatlantic exchange of plants, animals, cultural practices, technologies, diseases, ideas, and, of course, people between West Africa, Europe, and the "New World" (Crosby, 1972; Mann, 2005, 2011). The colonization and the respective death of ca. 56 million Indigenous people even led to a drop in the concentration of atmospheric CO_2, a human-driven impact on the Earth system, long before the industrial revolution (Koch et al., 2019). Increasing global mobility meant that humans were no longer tied to and forced to adapt to the ecozones surrounding them; instead, these ecozones became mixed when millions of people were transported around the globe against their will, and with them nonhumans of all kinds. These relocations occurred within short timeframes, which are incompatible with the processes of evolution taking part within cosmic timescales, causing many species to become extinct, for example, due to new opponents in the same ecological niche.

However, humans also invented technologies that could create novel species or revive extinct species. Since it is all too clear that humans have contributed to the Quaternary extinction, an increasing number of scientists call not only for more nature preservation, but also for a comprehensive renaturation and rewilding of the Earth, aiming to recreate a time when the human species did not exist or had no relevant influence on the Earth's systems (Donlan, 2005; Svenning et al., 2015).

In the case of the Pleistocene Park in East Siberia, the Pleistocene – the geological epoch lasting from 2.58 million to 11,700 years ago, preceding the Holocene – is currently re-created (Andersen, 2017; Macias-Fauria et al., 2020; Popov, 2020; Zimov, 2005). The evolutionary tables are turned here: humanity is not the child of its era, but humans create the era of their choice (Donlan et al., 2006). Thus, on 16 km^2 of Siberian tundra, a typical Pleistocenian wilderness is in the making, where herbivores like Yakutian horses, reindeer, moose, musk-oxen, bison, and yak are already living (see Figure 15).

The goal of the Pleistocene Park project is to revive a past ecosystem, which was destroyed by humans, in the hope that this will lead to more permafrost and thus less CO_2 and fewer methane emissions. First, the large herbivores are expected to trample snow into the ground; and second, the hoped-for conversion of tundra (back) to steppe is expected to result from an increase in solar reflection from the ground. This deep-time interaction thus interferes with the process of evolution by reaching into the past to alter the future.

The focus species for de-extinction is the mammoth, as a megafauna species which would profoundly shape a possible future Pleistocene grassland, with few but promising results to date: scientists were able to demonstrate the biological activity of a 28,000-year-old mammoth nuclei in mouse oocytes (Shapiro, 2015; Yamagata et al., 2019). More precisely, this "mammoth" is a rather new synthetic species called chimaera, in the form of a mix of mammoth and either Asian or African elephants as surrogate mothers. Thus, taking the assumed effects of this novel species into account, in the Pleistocene Park "synthetic biology meets geoengineering" (Herridge, 2021, p. 387). Even though human

Figure 15 Artist's impression of the landscape in the Pleistocene Park when the mammoth chimaera will be alive.

Source: © Raúl Martín Demingo (2013), available at www.raulmartin-paleoart.com/g2/g2/g2/g2/g2/g2/g2/g2/g2/g2/g2/g2/g3/1

societies have interfered with evolution for generations, this deep-time interaction could possibly kick-start a much more intense interference with much deeper effects on the Earth's biosphere, and probably geosphere in the long-term, if synthetic biology is applied widely within the coming decades.

Third, this touches upon the question of how to govern deep-time interactions that not only aim to bring back the landscapes of a past epoch, but to interfere intentionally and possibly on a large scale with the process of evolution in an irreversible manner, through synthetic biology.

Thus, this concerns the agency that artificially created parts of the biosphere and the influence it may exert in the long term, including, for example, the emergence of possible novel species, based on the "original creations" or natural dissemination across other parts of the world. Of course, as is the case with almost all new technology, the use of synthetic biology should not simply be declared good or bad as such, but, instead, should be recognized as a tool that can create effects in various ways and needs to be governed. Currently, the key organization striving to bring back the mammoth is a start-up company named Colossal, founded by a Harvard professor, the renowned geneticist, George Church, and the entrepreneur Ben Lamm. Considering the potential effects such a company might have, by creating and enabling a novel kind of agency to influence the process of evolution, and thus influence the course of life on planet Earth, the question regarding who should have a say in this process arises.

This leads directly to the question of democracy in relation to the political architecture of deep-time interactions in two ways. First, the public needs to be more closely involved in decisions that touch upon the very fundamentals of the process of evolution as a common good. In particular, local communities whose coexistence with local animals might be altered need to be involved, as these populations might mingle and mix their genes with chimaeras. Currently, the pace and irreversibility of these developments outpace legislation. The very first initiatives on regulating scientific advances have only just been formulated, ranging from rather broad groups like the Expert and Citizen Assessment of Science and Technology (ECAST) in the USA to more specialized groups like the Global Citizen's Assembly on Genome Editing.

Besides the involvement of a broader human public, experiments are needed to develop democratic and evidence-based inclusion of the process taking place within cosmic timescales, namely, evolution and its various (possible) parts, such as the genetic code of mammoth chimaeras that transfers genetic information across time. This is because processes, which have similar transformative potential with regard to planetary change as human societies, need to be represented democratically (see 3.2). Therefore, an observatory is needed to determine "how the potential of science can be better steered by the values and priorities of society" (Jasanoff & Hurlbut, 2018, p. 436). This should not only track and analyze research and real-world developments on biosynthetic biology and societal effects but should also mediate between both and identify possible ways of inclusion. If steered wisely, the democratization of synthetic biology might not slow down developments, and instead may provide additional input, work as a catalyst for innovation, and may anticipate potential risks and conflicts.

Of course, preventing extinction is the best way to make de-extinction arrangements at least partially redundant, just as mitigating climate change would have made adaptation redundant. For this reason, it is important to closely connect the politics of synthetic biology to networks that strive for similar goals with different means, such as the Legacy Landscape Fund, which aims to protect distinct landscapes rich in biodiversity or the Global Deal For Nature.

4.5 Space Weather Mitigation

First, deep-time interactions between societies and solar cycles, mediated by the magnetosphere, relate to the politics of deep time in terms of space weather mitigation. At its core, space weather mitigation concerns the management of the potential impacts of space weather events, such as solar flares and coronal mass ejections on various technological and societal systems, including power grids, satellite networks, and communications infrastructure. Since space weather events can have global impacts, establishing an international framework for data sharing and response coordination is essential. Another critical issue is the allocation of resources for space weather mitigation. Given the complexity and unpredictability of space weather events, it can be challenging to effectively allocate resources to prevent or minimize their impacts during the short period of time available after, for example, a major solar storm has been identified. Since space weather events are not as visible as other natural disasters, including hurricanes or earthquakes, they may not receive the same political attention and resources. Space weather mitigation policies also raise questions regarding accountability and responsibility. For example, if a space weather event causes significant disruption or damage, who is responsible for mitigation and recovery costs? Finally, the politics of mitigating space weather are tied to the broader debates about the politics of new technologies and the role of science in policymaking. As space weather events are a relatively new problem area, there is a need for continued research and scientific monitoring to better understand the impacts and to develop effective mitigation strategies.

Second, what are the main dynamics of the deep-time interaction? The solar cycle describes the change in the sun's activity based on the observation of sunspots on its surface, with the activity also manifesting in solar flares, coronal loops, and respective solar radiation. Solar cycles last on average 11 years and result from a magnetic reversal of the sun's poles, with maximum solar activity taking place during the reversal. While this solar activity has an effect on the Earth's climate and most likely human health, here the focus is on solar wind, particularly in the form of solar storms. They most likely, yet not necessarily, occur during the maximum period of solar activity in the 11-year-cycle (Paleari et al., 2022). Solar winds are part of cosmic rays, which also originate from the Milky Way and distant galaxies. Solar winds are a stream of charged particles, mainly protons, but also electrons and fully ionized atoms. Solar winds forming a solar storm that disturbs the Earth's magnetosphere, which usually exceeds the effects of normal solar winds, need to be considered as deep-time interactions of relevance for future societies.

If a solar storm, comparable to the so-called Carrington Event of 1859, which caused worldwide auroras as a result of the disturbance of the Earth's magnetosphere, hit Earth today, the damage would be considerable. It could potentially lead to a total breakdown of the technological infrastructure and a blackout of our power, navigation, and telecommunications systems, which would come at an estimated cost of over US$40 billion per day (Oughton et al., 2017).

Moreover, solar storms also have a direct influence on the human body, for example, causing higher rates of miscarriages as could be observed in cabin crew members (Baker, 2009; Belisheva, 2019; Grajewski et al., 2015; Lanzerotti, 2017; Riley et al., 2018; see Figure 16).

On February 4, 2022, for example, the company SpaceX lost forty satellites a day after its launch due to a geomagnetic storm. Moreover, the interruption of telecommunications could result in major geopolitical conflicts, such as on May 17, 1967, when the US Ballistic Missile Early Warning System broke down in several countries (Knipp et al., 2016). Bombers loaded with nuclear weapons were ready to make their way to the Soviet Union, as it was assumed that the communist regime was behind the incident, as it was preparing its own nuclear attack. Fortunately, the US Air Weather Service was able to clear up matters in time: the cause was, in fact, a solar flare, and it had not only interfered with the early warning system but also with the radio waves, so that it would have been impossible to recall the bombers once they were in the air. As societies increasingly rely on technology, the potential impact of solar storms on ever more aspects of human life also increases; in 2015, solar storms

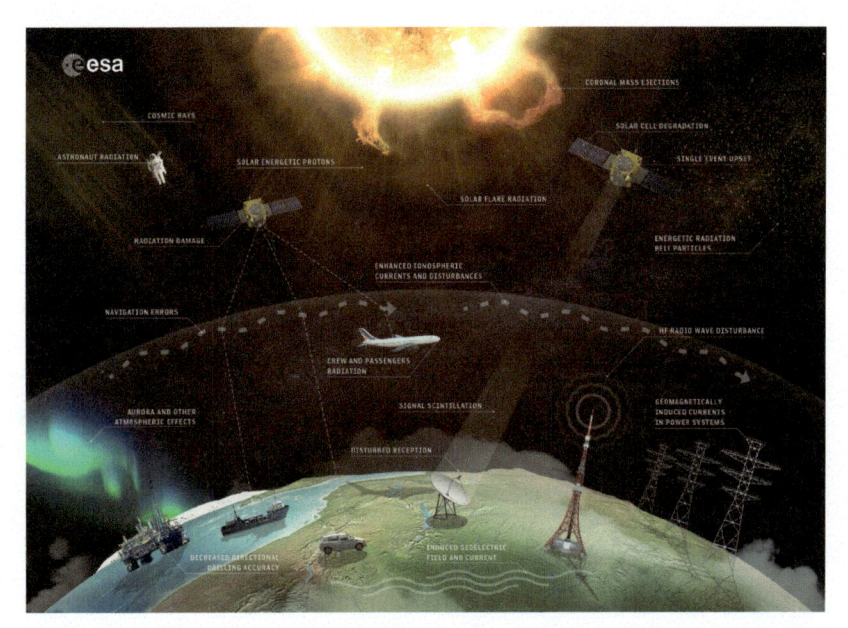

Figure 16 Solar winds and their influence on technology.
Source: © ESA/Science Office (2018), CC BY-SA 3.0 IGO, available at www.esa.int/ Space_in_Member_States/Germany/Weltraumwetter_Die_zerstoererische_Kraft_ der_Sonne

interfered with the global positioning systems in the United States and thus with self-driving cars. For how long (parts of) the world population would be without power, should a strong solar storm hit Earth, is yet unclear but estimations range from weeks to years.

Solar cycles and thus solar storms are likely to occur in the near future, but also constitute risks in the distant future until the end of the human species. In 2012, the probability of the occurrence of a Carrington-like event within the next decade was calculated at 12 percent (Riley, 2012). In that year, a solar storm missed the Earth's orbit by just nine days and would likely have cost US$10 trillion, as well as many lives, due to the subsequent technological disruptions (Baker et al., 2013). While solar cycles will thus influence all future human societies in one way or another and human societies may not be able to affect solar cycles, in turn, humanity may develop ways to mitigate and adapt to them. Since the mid-twentieth century, there has been unintentional anthropogenic space weather, caused in no small part by the long-term effects of atmospheric nuclear tests, such as Hardtack Teak in 1958 (Gombosi et al., 2017; Guglielmi & Zotov, 2007). However, human activity may also have a reverse effect on the magnetosphere: longitudinal waves used for communicating with submarines provide additional protection from the Van Allen radiation belt, a zone of highly charged and, therefore, deathly solar particles held at a distance from the Earth by the magnetosphere. This leads to the question of how human societies should shape their deep-time interactions with solar storms and, more broadly, how they should design the politics of deep time in relation to space weather mitigation. Unlike, for example, gene editing, which is already applied to deep-time interactions in the context of reviving the mammoth, the technology that could possibly become essential for space weather mitigation does not yet exist. To date, there are no ideas on how to strengthen the Earth's magnetosphere against a solar flare, how to manipulate the magnetosphere or whether, and how to govern such an attempt.

Second, from the perspective of solar storms and, more broadly, space weather mitigation as deep-time interactions, at least four elements for the respective politics of deep time can be identified: evidence, awareness, technology, and the democracy question. While the first two relate to the agency question, the latter two refer to the architecture.

The exact kind of agency of space weather and solar storms is, to a certain extent, still unknown. This includes the question of what processes within the Sun lead to major solar eruptions, which is why further explorations, such as the space probe, Solar Orbiter, run by the European Space Agency (ESA) in cooperation with the National Aeronautics and Space Administration (NASA), are crucial to develop evidence-based politics. The temporal unfolding of strong solar storms, in particular, makes their agency so dangerous: the probability of a Carrington-like solar storm hitting Earth is only 12 percent over a 10-year period, but if it is detected, societies would be likely to have less than a day, probably only 60 to 90 min, to protect and shut down technological infrastructure that would otherwise be damaged.

This leads to another aspect related to the agency of strong solar storms, namely, that it is necessary to raise awareness of their existence among the general public. This deep-time interaction may seem outlandish or highly speculative at first, but for the same reasons, most recommendations of virologists warning of a pandemic were brushed aside just a few years ago. Respective research roadmaps are calling for "a coordinated international approach to effectively provide awareness of the processes within the Sun–Earth system through observation-driven models" (Schrijver et al., 2015, p. 2746). So far, indications that humanity has but closely escaped such a catastrophe have largely been ignored, but more recently, the first policies have been introduced: In 2020, for example, the US House of Representatives passed and President Trump signed into law the "Promoting Research and Observations of Space Weather to Improve the Forecasting of Tomorrow Act" as a follow-up to a related strategy brought forward by the Obama administration in 2015. This kind of awareness raising among the broader public with regard to deep-time interactions can be described as the familiarization with a probable yet unknown future or, more precisely, the envisioning of events that only happen every few centuries (i.e., within cosmic timescales).

Political architectures are needed which enable the development and implementation of appropriate societal practices and technologies to cope with space weather. This includes building organizations. Of course, specialized institutions, such as the ESA's Space Weather Office, are still currently unable to properly manage potential deep-time interactions with strong solar storms, as their resources and equipment are vastly disproportionate to the possible magnitude of the interaction. In addition, to reach a broader audience, space weather forecasting needs to make its way into the daily lifeworld, such as becoming part of daily weather forecasts and, finally, becoming part of a planetary culture in which deep-time interactions are as recognized. Elders from different Indigenous groups can provide insights into this, as many Indigenous narratives and practices are rooted in the assumption that everything in the sky reflects everything on land and in the sea, and vice versa (Hamacher, 2022). While better forecasting is ultimately the best risk avoidance technique, the use of nonmagnetic steel in transformers and the increased use of surge protectors in the grid are also approaches which are already available to build resistance against the impacts of space weather. However, considerable interventions have also been proposed, including a gigantic space shield located between Earth and the Sun (Lingam & Loeb, 2017). The shield, which resembles an electrified copper loop, would have to be nearly as large as the Earth, would weigh 105 t and cost an estimated US$100 billion. The questions that arise with such gigantic infrastructures are not only whether they are feasible, which they could be at some point in the future, but also whether, how, and by whom they could be governed. This requires a reflexive element in the politics of deep-time interactions, similar to technological impact assessments, yet extended to include what might be called a "governability impact assessment."

Lastly, the nature of societal interference with space weather and the respective forecasting in combination with novel technologies can potentially lead to a technocracy. On the contrary, particularly with regard to the "planning" of deep-time interactions, it becomes clear that this has to be done with great emphasis on democratic principles. If the right to an unprojected future is seen as a core normative deep-time principle, it would have been to be protected as an inalienable right, which means that even political arrangements with regard to a specific deep-time interaction must be able to be re-designed by future generations. Perhaps future generations will find ways to of making productive use of solar storms, requiring a 180-degree turn in their respective political arrangements, which would most likely focus initially on resilience measures. This is where the trustee conception of sovereignty assumes upmost importance: trustees of a democratic process need to guarantee that this process takes place across all generations, allowing future generations to build and shape deep-time interactions with solar storms as they see fit.

4.6 Terraforming Mars

First, planet reformation, here the potential terraforming of Mars, that is, purposefully altering the planet's environment to make it more livable for human life, is inherently a political issue. Just a few centuries ago, humans landing on the moon was unimaginable, which is why the possibility of humans on Mars should be taken seriously when considering possible deep-time interactions. What once used to be part of science fiction, namely, a multiplanetary human society, does not seem fundamentally impossible anymore, as there is currently a renaissance of government and private spaceflight, which must be researched, operated, and evaluated from the outset as a deep-time interaction. While a wide range of policy areas have been analyzed by Earth system governance scholars, research with respect to spaceflight is still rare (Ferretti, 2021; Newman & Williamson, 2018; Suchantke et al., 2020). The decision to terraform Mars involves complex ethical and political considerations, including issues of environmental impact, resource allocation, and power dynamics. Proponents argue that terraforming Mars could create a new frontier for human expansion and reduce the risk of human extinction by creating a contingency plan for humanity. However, critics point out that the resources required for terraforming could be better spent on solving pressing social and environmental problems on Earth, and that the risks of irreversible ecological damage or harm to potentially Indigenous life forms on Mars are too high. Attempting to reshape an entire planet would require significant resources, expertise, and collaboration among governments, corporations, and other entities. This raises questions about who would have access to the benefits of terraforming, who would bear the costs and who would have control over the process. In addition, the colonization of Mars raises questions about imperialism, neocolonialism, and the ethics of colonizing a new world. The politics of terraforming Mars therefore involve complex tradeoffs and negotiations among various stakeholders, including scientists, policymakers, entrepreneurs, and the public, and require careful consideration of the ethical and political implications of human expansion beyond Earth. The deep-time interaction between societies and planet (re)formation thus relates to the politics of deep time in outer space over extremely long timescales.

Second, what are the core dynamics regarding a possible terraforming of Mars and which political arrangements – on Mars and on Earth – are required to prevent, shape or even foster this? When considering any form of Mars terraforming as a possible deep-time interaction, its feasibility needs to be inspected. Private companies are currently investing heavily in space travel, making it accessible at a lower cost. With this tailwind, major government-funded space agencies are likewise refocusing their attention on destinations beyond the Earth's orbit, which is, to some extent, also part of their geopolitical strategy. In recent years, the Moon has once again become a target for human spaceflight, and thereafter, the next target is likely to be Mars. Thus, looking beyond space tourism by billionaires, serious approaches of human missions to Mars and their respective stations or even settlements seem to be only a matter of time: this is not only due to the aspirations of nation states in the so-called "new space age," especially on the part of the United States and China – China aims to build an ultra-large spacecraft more than a kilometer in length – but also due to private business investments which have accelerated cost reductions in space travel (Capova, 2016; Chik, 2021). In particular, the cost per unit of mass transported from Earth into outer space is becoming increasingly cheaper: launching a space shuttle into orbit in the 1980s cost over US$50,000 kg^{-1}, whereas the estimates for the SpaceX' Starship, which is scheduled to launch in the 2020s, are "well under US$100 kg^{-1}" (Musk, 2020; see Figure 17). Reusable launch systems, in

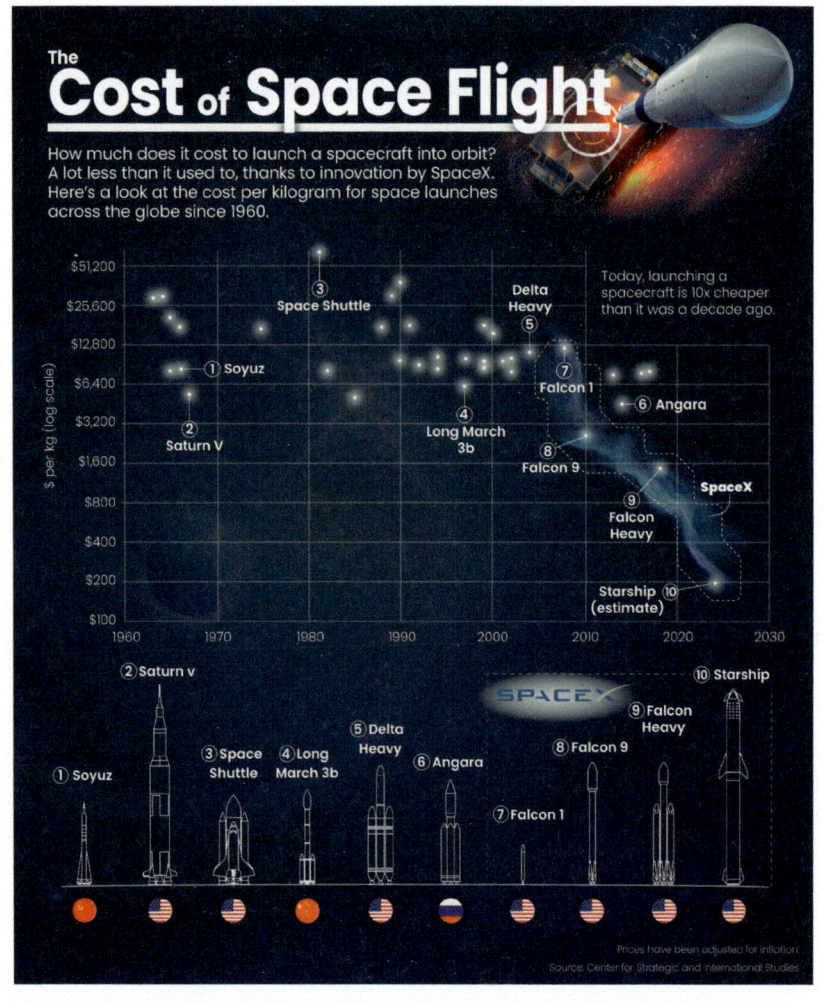

Figure 17 Cost reduction of space flight since 1960.

Source: © Visual Capitalist/Bruno Venditti and Sam Parker (2022), available at www.visualcapitalist.com/the-cost-of-space-flight/

particular, are crucial in this development, and future technological breakthroughs are also possible, such as the production of methane in outer space as fuel for rockets. In other words, the temporal feature of acceleration can cause deep-time interactions to become feasible in much shorter timeframes than one could expect, based on current and past developments.

The main reason for managing the terraforming of Mars as a deep-time interaction is, of course, its timescale. Planet reformation and the development of habitable conditions for complex life, based on the only case known so far, requires millions and billions of years. Terraforming an already existent planet, which, to the best of our knowledge, has never previously been done, not only means interfering with and initiating processes taking place within cosmic timescales in the distant future, but in the case of Mars, these plans are largely aimed at accelerating these processes and compressing the timescale. The techniques and timelines proposed to terraform Mars are manifold, pointing to many unresolved implementation challenges (Beech et al., 2021; Green et al., 2021). One of these timelines, for example, proposes two consecutive steps after a pre-terraforming phase. The purpose of the pre-terraforming phase is to collect more information regarding conditions on Mars, ranging from an inventory of the number of certain chemical elements and their abundance to the question of whether life exists on Mars and whether it should be interfered with (Berliner & McKay, 2017). Subsequently, the first step of terraforming is a warming phase of ca. 100 years, which aims to heat up Mars' surface from −60 °C to +15 °C, by releasing CO_2 from Mars' polar caps and greenhouse gases produced by humans. The second step is an oxygenation phase of ca. 100,000 years, which will probably require the use of synthetic biology and biotechnology since photosynthesis is unlikely to work on (all of) Mars. Here, the bidirectionality of deep-time interactions becomes apparent. On the one hand, and regardless of the feasibility, the notion of a possible terraforming of Mars some 100,000 years from now is not only driving the technological developments of outer space exploration, which can be used for purposes as diverse as monitoring geodynamics and climate change through satellite geodesy and advancing the military buildup of, for example, the Unites States Space Force, but is also shaping cultural narratives of space as the last frontier. On the other hand, decisions here and now, ranging from non-terraforming agreements to international collaborations to promote terraforming on Mars, can influence and shape the future of Mars in the next hundreds of thousands of years and beyond.

Whether terraforming is feasible is still an open question, since the complex dynamics of the Earth system, that developed over billions of years, are obviously difficult to reproduce, as the real-world experiment, "Biosphere 2," has demonstrated. The research facility, "Biosphere 2," a closed miniature ecosystem in the Arizona desert, was used in the 1990s to examine Earth system processes on a small scale, not least with regard to the need for negative emissions. "Engineered to be a self-sustaining mesocosm" (Avise, 1994, p. 327) – that might one day be used for extraterrestrial purposes – Biosphere 2 boasted a number of different "natural" habitats: "tropical rainforest, marsh, desert, savannah, streams, agricultural area, and even a miniature ocean complete with coral reef" (Avise, 1994, p. 327). There were two long-term experiments involving human settlers: one from 1991 to 1993 and another in 1994 (Nelson, 2017, 2018). Both crews failed to operate Biosphere 2 without outside help. The main problems were that the concrete used for construction reacted with carbon dioxide, binding oxygen in the process, and that the microbes on the artificial field enriched the atmosphere with too much nitrogen and carbon dioxide. What is more, "the miniaturization enormously accelerated biological cycles" (Rand et al., 2016/2021). Finally, there were conflicts among the crew, involving psychological, social and political aspects, in addition to what was happening in the "natural" systems. In addition, Martian settlers would have to understand not only the Earth system, but also the Mars system, and there would be many open questions to address, such as how to feed larger populations of Martian settlers (Cannon & Britt, 2019). Thus, more generally speaking, deep-time interactions might require manifold iterations and feedback-loops in order to function and be shaped in a certain direction. This means that a continuous and institutionalized process of monitoring, assessment, and adjustment would help understand these interactions and would steer them in the desired direction.

Third, even though Mars will not be terraformed by the end of this century, a self-sustaining city on Mars is scheduled by SpaceX to be created within this timeframe. Therefore, the issue of politics is more pressing than terraforming timescales imply. Rather than envisioning a specific Martian government, this involves the creation of conditions for a trustee process of sovereignty that might, for example, allow a possible future Martian population to write its own constitution. Thus, politics relating to deep-time interactions with terraforming and the colonization of Mars can be separated into the terraforming politics on Earth and, later, the politics of Mars on Mars.

The agency that the terraforming of Mars currently exerts barely goes beyond the aspect of imagination. However, depending on whether and how the formation of respective visions aligns with the deep-time normative core principles of habitability and democracy, corresponding path dependencies may unfold. Once established, realities on Mars ranging from power structures to evolutionary processes, which potentially start to evolve with the first Mars station, may be nearly impossible to reverse. Thus, the potential terraforming of Mars creates an agency that requires respective capacity building in the here and now. On Earth, capacity building initiatives for a political architecture, which supports the realization of habitability on Mars, started rather implicitly with the foundation of the United Nations Committee on the Peaceful Uses of Outer Space (COPUOS) in 1959. However, in the context of the United Nations' Sustainable Development Goals, particularly the "Space2030 Agenda," habitability initiatives have accelerated (Ferretti, 2021): the COPUOS established a working group on the "Space2030 Agenda" in 2018, and other sustainability activities are being implemented under the heading "Long-term Sustainability of Outer Space Activities," including the United Nations' "COPUOS Guidelines on Long-term Sustainability of Outer Space Activities," which were adopted in 2019. The ESA, to give another example, already engages in a broad range of activities related to the 17 Sustainable Development Goals (https://sdg.esa.int). Whether and how an extension and transfer of these activities pertaining to planet Earth can be applied to the potential terraforming of Mars is still an open question. Again, the potential terraforming of Mars, for example, by private companies, requires that the politics of deep-time interactions take into account processes taking place within cosmic timescales in the here and now.

Questions remain regarding the way in which deep-time interactions, related to the terraforming of Mars, can be realized in a democratic manner, particularly the way in which the sovereignty of any future Martian population can be guaranteed and constantly renewed. What would be the outcome should Mars' population declare independence one day? Of course, the Outer Space Treaty of 1967 acknowledges negative freedom by stating that "Outer space, including the moon and other celestial bodies, is not subject to national appropriation by claim of sovereignty, by means of use or occupation, or by any other means" (Article II). However, the current legal framework is incomplete, as it does not provide any guidance on the way in which the notions of sovereignty could govern a process of terraforming and government formation on Mars (Leib, 2015; Van Eijk, 2020; Wójtowicz & Szocik, 2021). If viewed as an interplanetary common, the colonization, terraforming, and potential exploitation of Mars, not only as a deep-time interaction but also as a deep-space interaction, need to be regulated or better prepared than, for example, current deep-sea mining activities. Otherwise, single states or companies might move forward without the appropriate legislation in place. To illustrate this, I once again use the example of SpaceX, as it is the company most likely to conduct a manned mission to Mars and has already included a section on a future Martian government in the terms and conditions of its satellite-based internet service, Starlink: "For Services provided on Mars, or in transit to Mars via Starship or other colonization spacecraft, the parties recognize Mars as a free planet and that no Earth-based government has authority or sovereignty over Martian activities. Accordingly, Disputes will be settled through self-governing principles, established in good faith, at the time of Martian settlement" (SpaceX Starlink, 2020). Therefore, it becomes clear that deep-time political architectures need to be established in the here and now to enable a process that is not dominated by individual interests, even if the eventual terraforming of Mars will not be completed for another 100,000 years or so.

5 A Conceptual Framework of the Politics of Deep Time

> The human mind may not have evolved enough to be able to comprehend deep time, it may only be able to measure it.
>
> (McPhee, 1981, p. 127)

No politics will last one million years, not even 100,000, 10,000 or 1,000 years. Yet, as I have shown, human societies in the realm of deep time interact with processes taking place within a geological or even cosmic timescale. In other words, no political arrangement can be expected to last as long as the epoch-spanning processes that it deals with. The fewer deep-time political structures that need to be kept in place, the better. These are hard to maintain because there is no guarantee that they will outlive the next major war, that they will be shut down due to financial restrictions or that they will slowly die and be forgotten. Therefore, the challenge for the politics of deep time is that it must function not only across various levels, but also across various eras, and yet, must always remain situated in the here and now.

I propose the following overview of the politics of deep time as both a conceptual framework and a political necessity (see Figure 18). Of course, this overview is based on the insights that result from my pioneering study of the politics of deep time and might be adjusted based on the broader empirical coverage of deep-time interactions and their respective deep-time politics.

While many nuances exist in each of the cases investigated, two main characteristics of deep-time interactions have become apparent to varying degrees across all of them: irreversibility and uncertainty.

irreversibility & uncertainty
of deep-time interactions

require

deep-time observatory
for deep-time literacy

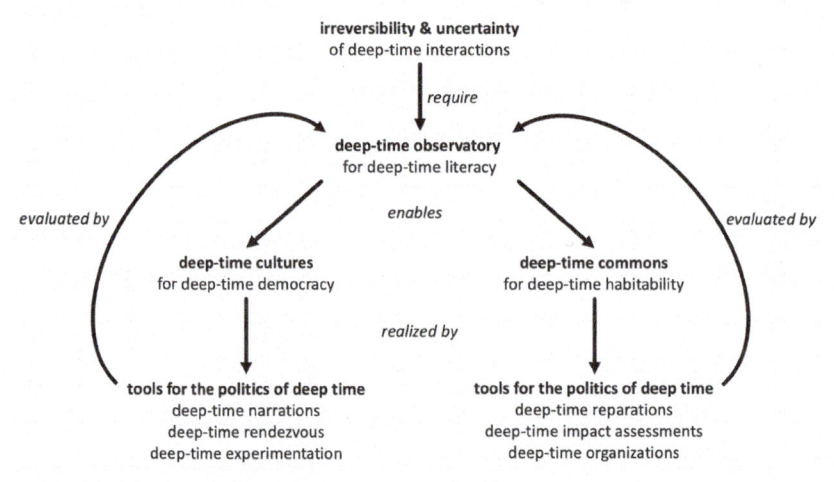

evaluated by *enables* *evaluated by*

deep-time cultures
for deep-time democracy

deep-time commons
for deep-time habitability

realized by

tools for the politics of deep time
deep-time narrations
deep-time rendezvous
deep-time experimentation

tools for the politics of deep time
deep-time reparations
deep-time impact assessments
deep-time organizations

Figure 18 Organic emergence and functioning of the politics of deep time

In terms of irreversibility, I have observed, for example, that if karst aquifers are destroyed, they cannot be replaced, or that the terraforming of Mars, once initiated, is nearly impossible to undo and instead takes on a life of its own. Irreversibility thus refers to the fact that once certain processes have been set in motion over geological or cosmic timescales, it is impossible to reverse or undo their effects on human and nonhuman societies. For instance, the burning of fossil fuels and the subsequent emission of greenhouse gases have already set in motion irreversible changes to the Earth's systems. The scale and pace of such undoable changes make it difficult to predict the extent and nature of their impacts on contemporary and future societies.

Uncertainty has been observed, for example, regarding the effects a future solar storm might have on Earth, or the kind of symbols that can transport knowledge of a final depository for nuclear waste across one million years and numerous generations. In other words, not only are there deep-time interactions that lead to irreversible changes, but vast knowledge gaps also exist regarding these interactions, some details of which will probably never be known, despite the best attempts in research to do so, such as the Deep-Time Digital Earth program (Wang et al., 2021). Uncertainty thus arises from the fact that deep-time interactions involve complex planetary systems and processes operating over long periods of time. This complicates the prediction of the future behavior of these with any degree of accuracy, which, in turn, makes it difficult to plan and take decisions with confidence. This also means that, due to the degree of unpredictability unexpected, unforeseen events may occur, which can have significant and sometimes catastrophic consequences for human and nonhuman societies.

Together, irreversibility and uncertainty form a dangerous combination if not handled wisely. This underscores the importance of developing a deep-time perspective and the respective intentional politics of deep time when considering human interactions with geologic and cosmic processes. Such a perspective requires that the limits of knowledge and predictive ability be recognized and the need for caution and humility are acknowledged. This can help to better understand the risks and opportunities associated with deep-time interactions.

My proposal for a deep-time observatory comes in at this point. A deep-time observatory can function as a competence center for deep-time literacy in the here and now, which will benefit not only decision-makers but all human and more-than-human societies. Such an observatory may be composed of one-third representatives of the humanities, social sciences, and natural sciences; one-third representatives of human societies; and one-third representatives of nonhumans or even more purposeful with the help of the respective technology-enabled "voices," allowing through their composition an extended peer review (Funtowicz & Ravetz, 2001). One of the first major tasks of this observatory would be to compose an inventory of deep-time interactions and the associated bidirectional influences between societies in the here and now and processes on cosmic timescales.

First, the deep-time observatory enables deep-time cultures that in turn enable deep-time democracy, which can take different forms, depending on the deep-time interaction they address. Second, the information produced by a deep-time observatory provides the foundations for identifying and protecting deep-time commons – understood as resources, spaces, or knowledge that exist and function within a geological or cosmic timescale and are shared by multiple entities, including human and nonhuman actors – to ensure deep-time habitability. Such protective measures could entail declaring karst aquifers a World Heritage Site or classifying landscapes with rich biodiversity, such as Legacy Landscapes.

Deep-time culture refers to the recognition and valuing of the interconnectedness of all life and ecosystems within a geologic timescale. It recognizes the interference of human actions with the planet over large timescales and seeks to develop practices that address their democratic governing. A deep-time culture is characterized by a long-term perspective and an understanding that humans are part of the planet, not separate from it.

Deep-time cultures are realized by respective tools. These include understanding old narrations and formulating new ones, which can take the form of evidence-based myths, as in the case of karst aquifers or symbols which help to remember, for example, the power of solar storms once knowledge of them may have been forgotten after a civilization collapse. Such deep-time myths can be thought of as narratives that depict the connections between human societies and processes taking place within geological or cosmic timescales, in order to shed light on the way in which human societies have understood their place on this planet and how they have interacted with nonhuman forces over long periods of time. As part of the politics of deep time, deep-time myths could play a crucial role in shaping collective imagination and guiding actions. Based on scientific data and models, new deep-time myths can be developed to transport experiences and knowledge of deep-time interactions. As part of the politics of deep time, such myths could be used to influence public discourse and policymaking, be incorporated into curricula, and used for public awareness campaigns. Myths of this kind are then similar to scientific records (Nunn, 2018, 2021).

A second tool is the deep-time rendezvous, as this is needed for deep-time interactions that may require reassessment from time to time, such as the final depositories for nuclear waste. Deep-time rendezvous needs to include remembrance and thus some form of durable communication, such as establishing protocols for preserving and sharing knowledge across generations and across cultural and linguistic boundaries. Another example would be the establishment of a deep-time park or reserve that protects and preserves areas significant for their deep-time interactions, providing opportunities for people to learn more about these interactions, such as the Deep Time Walks (www.deeptimewalk.org). Moreover, deep-time monuments may be developed that allow to connect the history of humans and Earths in a way that they transcendent geological epochs by respective semiotics (Szerszynski, 2017). Ultimately, the continuity of deep-time rendezvous would have to be ensured by making them valuable and meaningful for future generations.

Deep-time experiments constitute a third tool, which is particularly useful in finding ways to establish trustee conceptions of sovereignty, so that democracy exists through deep time, for example, in the potentially long process of terraforming and establishing a government on Mars. Deep-time experiments are thus a way of exploring alternative social and political practices that take into account the temporal dimensions of more-than-human societies in democratic politics, similar to the Embassy of the North Sea (www.embassyofthenorthsea.com). The aim must be to identify which institutions – not necessarily centered on a parliament – and processes – also designed for nonhumans unable to communicate orally or in a written form – have to be invented. Experimentation can make use of the potential of sensors, machine learning, and semiotics that allow humans to understand the "voice" of processes taking place within geological and cosmic timescales. Respective technologies can be used from the inner Earth to interplanetary space in order to identify the signs and meaning of more-than-human agencies across various spheres, such as volcanoes as an expression of slowly melting rocks that press to the outside.

Deep-time commons refer to a resource or feature of the planet that develops within geological or cosmic timescales, has existed over a long period of time, and will continue to exist and interact with multiple generations. This may not only be a fruitful soil as the Black Belt, but also a geological formation, such as a mountain or canyon, which has existed for millions of years and is interfering with human societies, possibly also being valued for its beauty and cultural significance. Another example is a body of water, such as a river or lake, which has hosted human communities and ecosystems for centuries or even millennia. Deep-time commons are put into effect using the appropriate political tools.

These include deep-time reparations to provide a novel future of choice for those who have been hurt by the misuse of resources through deep-time interactions, such as African Americans in Alabama. Deep-time reparations recognize that the actions of certain human societies have had, and will have, long-lasting and often irreversible impacts on the planet, which have often disproportionately affected marginalized communities and future generations. Deep-time reparations thus represent a shift from the idea of traditional reparations that focus on addressing past harms. This could include efforts to restore ecosystems or habitats and provide support for communities affected by long-term degradation. Importantly, these actions would need to be forward-looking, aimed at preventing further harm yet opening up new possibilities enabling the affected communities to thrive.

Another tool is the execution of deep-time impact assessments that investigate the way in which novel technologies, such as synthetic biology that creates chimaeras, can manipulate deep-time interactions and thus alter deep-time commons either positively or negatively. As such, technologies alter the genetic composition of organisms; they might have significant impacts on the biodiversity of nonhuman societies and the biogeochemical cycles of the planet, with effects on ecosystems or food webs. In a similar manner, the impact of mining or drilling over long periods of time could be assessed, as well as the potential impact of natural events, such as meteorite impacts on human society over centuries or even millennia, including feedback loops and the consideration of unexpected consequences that might occur.

Yet another tool is the establishment of deep-time organizations that deal with specific deep-time interactions, such as the Svalbard Global Seed Vault safeguarding the genetic diversity of crops. Such deep-time organizations would have to exist over very long periods of time to deal with deep-time interactions, and for this reason, their institutional setup is of crucial importance. It has been shown that distinct design principles can be identified when establishing these organizations, such as placing the organization in a safe area while ensuring its societal embeddedness, focusing on one core place of outreach instead of diversification, or creating ownership for the public in the decision-making of the organization (Hanusch & Biermann, 2020).

Of course, the aforementioned instruments are examples, and more tools may be developed and applied to realize democracy and habitability on and potentially beyond planet Earth in the realm in which societies interact with processes taking place within a geological and cosmic timescale. The tools are subject to evaluation by the deep-time observatory; in an iterative and reflective process the observatory may adapt its inventories, which could possibly lead to novel insights into how to enable deep-time commons and a deep-time culture. This means that the deep-time observatory would serve as a tool for monitoring and inventorying the interactions between societies and the geologic and cosmic processes that shape our planet and universe over long periods of time. This would involve the development of a framework for collecting and organizing data on deep-time interactions, which could include the use of standardized methods for data collection and analysis, and the development of a central database or platform for storing and sharing data. As part of this process, the deep-time observatory would assess existing practices and policies to determine their long-term impact, sharing best and worst practices and collaborating on solutions to deep-time challenges. In addition, it would need to identify practices so that data could be passed across technological and human generations (Jarvenpaa & Essén, 2023; Szerszynski, 2020).

Currently, the politics of deep time are scattered. These may grow organically, but targeted research, as well as tangible policy, may require an initial impulse. While a new constitutional moment that recognizes deep-time interactions across international, national, and regional political organizations is rather unlikely, I propose an alternative. This alternative is the creation of a deep-time observatory, which can even be tested as a prototype over a number of years and evaluated and refined. Of course, from the very beginning it should seek close interchange with political practitioners.

Practically speaking, the establishment of a deep-time observatory as a first step toward the intentional politics of deep time would be a largely altruistic act, as changes in societal relations, involving processes taking place within geological or cosmic timescales, would probably exceed election periods of a few years. However, a starting point leading to the realization of a deep-time observatory could be to build the very first prototype, by explicating its importance to decision-makers and the broader public, building a coalition of international allies, and developing a plan for its operation.

Whether the politics of deep time is the starting point for the broader politics of planetary times or whether the former makes its way into existing politics is, as of now, an open question. The purpose of any politics of deep time is far from being clear, especially when we think about the very distant future in which uncertainty exists as to the interests of any future beings. The case of species extinction illustrates this problem well. The considerable history of the Earth teaches us that major extinction events, while disastrous for incumbent forms of life, are followed by huge boosts in biodiversity and macroevolutionary innovation, such as the end of the dinosaurs and the rise of the mammals. How do we balance the imperative to restore what the planet once could do and conserve what the planet can do now, against the imperative to be open to the as-yet unrealized potential of the planet to embark on radically new projects in the future? I hope this pioneering proposal may function as a first step into the politics that strive to search for answers in this regard, opening up possibilities for alliances that seek to better understand and realize deep-time habitability and democracy.

References

Adam, B. (1998). *Timescapes of Modernity: The Environment and Invisible Hazards*. Routledge. https://doi.org/10.4324/9780203981382.

Adam, B., & Groves, C. (2007). *Future Matters: Action, Knowledge, Ethics*. Brill.

Andermann, T., Faurby, S., Turvey, S. T., Antonelli, A., & Silvestro, D. (2020). The Past and Future Human Impact on Mammalian Diversity. *Science Advances*, *6*(36), eabb2313. https://doi.org/10.1126/sciadv.abb2313.

Andersen, R. (2017, April). Welcome to Pleistocene Park. *The Atlantic*. www.theatlantic.com/magazine/archive/2017/04/pleistocene-park/517779/.

Antonello, A., & Carey, M. (2017). Ice Cores and the Temporalities of the Global Environment. *Environmental Humanities*, *9*(2), 181–203. https://doi.org/10.1215/22011919-4215202.

Archer, D. (2009). *The Long Thaw: How Humans Are Changing the Next 100,000 Years of Earth's Climate*. Princeton University Press. https://doi.org/10.2307/j.ctvct0042.

Arendt, H. (1958). *The Human Condition*. University of Chicago Press.

Arendt, H. (1993). Was ist Politik. In U. Ludz (Ed.), *Was ist Politik?: Fragmente aus dem Nachlaß* (1st ed., pp. 9–12). Piper Taschenbuch. (Original work published 1950).

Avise, J. C. (1994). The Real Message from Biosphere 2. *Conservation Biology*, *8*(2), 327–329. https://doi.org/10.1046/j.1523-1739.1994.08020327.x.

Back, W. (1981). Hydromythology and Ethnohydrology in the New World. *Water Resources Research*, *17*(2), 257–287. https://doi.org/10.1029/WR017i002p00257.

Bai, X., van der Leeuw, S., O'Brien, K. et al. (2016). Plausible and Desirable Futures in the Anthropocene: A New Research Agenda. *Global Environmental Change*, *39*, 351–362. https://doi.org/10.1016/j.gloenvcha.2015.09.017.

Baker, D. N. (2009). What Does Space Weather Cost Modern Societies? *Space Weather*, *7*(2), 1–2. https://doi.org/10.1029/2009SW000465.

Baker, D. N., Li, X., Pulkkinen, A. et al. (2013). A Major Solar Eruptive Event in July 2012: Defining Extreme Space Weather Scenarios. *Space Weather*, *11*(10), 585–591. https://doi.org/10.1002/swe.20097.

Bakke, M. (2017). Art and Metabolic Force in Deep Time Environments. *Environmental Philosophy*, *14*(1), 41–59. https://doi.org/10.5840/envirophil20173744.

Bakker, K. (2022). *The Sounds of Life: How Digital Technology Is Bringing Us Closer to the Worlds of Animals and Plants.* Princeton University Press.

Barad, K. (2007). Meeting the Universe Halfway: Quantum Physics and the Entanglement of Matter and Meaning. Duke University Press. https://doi.org/10.1515/9780822388128.

Barca, S. (2020). *Forces of Reproduction: Notes for a Counter–Hegemonic Anthropocene.* Cambridge University Press. https://doi.org/10.1017/9781108878371.

Battersby, P. (2017). Organizations and Globalization. In A. Farazmand (Ed.), *Global Encyclopedia of Public Administration, Public Policy, and Governance* (pp. 1–15). Springer International. https://doi.org/10.1007/978-3-319-31816-5_1226-1.

Bauer, A. M., Edgeworth, M., Edwards, L. E. et al. (2021). Anthropocene: Event or Epoch? *Nature, 597,* 332. https://doi.org/10.1038/d41586-021-02448-z.

Beach, T., Luzzadder-Beach, S., Cook, D. et al. (2015). Ancient Maya Impacts on the Earth's Surface: An Early Anthropocene Analog? *Quaternary Science Reviews, 124,* 1–30. https://doi.org/10.1016/j.quascirev.2015.05.028.

Beck, P. U. (1992). *Risk Society: Towards a New Modernity* (M. Ritter, Trans.). Sage.

Beech, M., Seckbach, J., & Gordon, R. (Eds.) (2021). *Terraforming Mars* (1st ed.). Wiley Scrivener. https://doi.org/10.1002/9781119761990.

Belisheva, N. K. (2019). The Effect of Space Weather on Human Body at the Spitsbergen Archipelago. In M. Kanao, G. Toyokuni, & Y. Kakinami (Eds.), *Arctic Studies: A Proxy for Climate Change* (pp. 55–70). IntechOpen.

Benda, H. J., & Castles, L. (1969). The Samin Movement. *Bijdragen Tot de Taal-, Land- En Volkenkunde, 125*(2), 207–240.

Bendell, J. (2018). *Deep Adaptation: A Map for Navigating Climate Tragedy* (Vol. 2) [Institute for Leadership and Sustainability (IFLAS) Occasional Papers]. www.lifeworth.com/deepadaptation.pdf.

Berlin, I. (1969). Two Concepts of Liberty. In I. Berlin (ed.), *Four Essays on Liberty* (pp. 118–172). Oxford University Press.

Berliner, A. J., & McKay, C. P. (2017, February 27). *The Terraforming Timeline.* Planetary Science Vision 2050 Workshop, Washington, DC.

Biermann, F. (2021). The Future of "Environmental" Policy in the Anthropocene: Time for a Paradigm Shift. *Environmental Politics, 30*(1–2), 61–80. https://doi.org/10.1080/09644016.2020.1846958.

Bjornerud, M. (2018). *Timefulness: How Thinking Like a Geologist Can Help Save the World.* Princeton University Press. https://doi.org/10.23943/9780691184531.

Blaser, M. J. (2014). *Missing Microbes: How the Overuse of Antibiotics Is Fueling Our Modern Plagues*. Henry Holt.

Bobbette, A., & Donovan, A. (Eds.) (2019). *Political Geology: Active Stratigraphies and the Making of Life*. Palgrave Macmillan.

Bonneuil, C., & Fressoz, J.-B. (2016). *The Shock of the Anthropocene: The Earth, History and Us* (D. Fernbach, Trans.; Reprint). Verso Books. (Original work published 2013).

Boston, J. (2016). *Governing for the Future: Designing Democratic Institutions for a Better Tomorrow*. Emerald Group.

Bostrom, N. (2013). Existential Risk Prevention as Global Priority. *Global Policy*, 4(1), 15–31. https://doi.org/10.1111/1758-5899.12002.

Brand, S. (Ed.) (1968). *Whole Earth Catalog*. Portola Institute.

Brand, S. (1999). The Clock of the Long Now. Time and Responsibility. New York: Basic Books. Page 164.

Brand, S. (2009). *Whole Earth Discipline*. Viking Penguin.

Braudel, F. (1966). *La Méditerranée et le monde méditerranéen à l'époque de Philippe II 2 Bde*. Armand Colin.

Bridle, J. (2022). *Ways of Being: Beyond Human Intelligence*. Allen Lane.

Brosig, A., Bräutigam, B., Barth, A., & Stanek, K. P. (2020). *Evaluierung des Kenntnisstandes von aktiven Störungszonen in Deutschland (KaStör)* (Forschungsberichte zur Sicherheit der nuklearen Entsorgung) [Abschlussbericht]. Bundesamt für die Sicherheit der nuklearen Entsorgung (BASE). www.base.bund.de/SharedDocs/Downloads/BASE/DE/fachinfo/fa/KaStoer_Abschlussbericht_2020.pdf?__blob=publicationFile&v=4.

Brumm, A., Oktaviana, A. A., Burhan, B. et al. (2021). Oldest cave art found in Sulawesi. *Science Advances*, 7(3), eabd4648. https://doi.org/10.1126/sciadv.abd4648.

Brunnengräber, A., & Di Nucci, M. R. (Eds.) (2019). *Conflicts, Participation and Acceptability in Nuclear Waste Governance: An International Comparison Volume III*. Springer VS. https://doi.org/10.1007/978-3-658-27107-7.

Brunnengräber, A., Di Nucci, M. R., Losada, A. M. I., Mez, L., & Schreurs, M. A. (Eds.) (2015). *Nuclear Waste Governance: An International Comparison*. Springer.

Brunnengräber, A., Di Nucci, M. R., Losada, A. M. I., Mez, L., & Schreurs, M. A. (Eds.) (2018). *Challenges of Nuclear Waste Governance: An International Comparison Volume II*. Springer.

Buckland, A. (2013). Novel Science: Fiction and the Invention of Nineteenth-Century Geology. University of Chicago Press. https://doi.org/10.7208/9780226923635.

Bundesamt für die Sicherheit der nuklearen Versorgung (BASE). (2022a, February 8). *Tasks Performed by BASE* [Internetauftritt einer Bundesbehörde]. www.base.bund.de/EN/bfe/tasks/tasks_node.html.

Bundesamt für die Sicherheit der nuklearen Versorgung (BASE). (2022b, February 28). *Übersicht der laufenden Projekte (Overview over Ongoing Projects)* [Internetauftritt einer Bundesbehörde]. www.base.bund.de/DE/themen/fa/soa/projekte-aktuell/projekte-aktuell.html.

Burchfield, J. D. (1998). The Age of the Earth and the Invention of Geological Time. In D. J. Blundell & A. C. Scott (Eds.), *Lyell: The Past Is the Key to the Present* (pp. 137–143). Geological Society.

Butterfield, N. J. (2011). Animals and the Invention of the Phanerozoic Earth System. *Trends in Ecology & Evolution*, *26*(2), 81–87. https://doi.org/10.1016/j.tree.2010.11.012.

Canfield, D. E., Glazer, A. N., & Falkowski, P. G. (2010). The Evolution and Future of Earth's Nitrogen Cycle. *Science*, *330*(6001), 192–196. https://doi.org/10.1126/science.1186120.

Cannon, K. M., & Britt, D. T. (2019). Feeding One Million People on Mars. *New Space*, *7*(4), 245–254. https://doi.org/10.1089/space.2019.0018.

Capova, K. A. (2016). The New Space Age in the Making: Emergence of Exo-Mining, Exo-Burials and Exo-Marketing. *International Journal of Astrobiology*, *15*(4), 307–310. https://doi.org/10.1017/S1473550416000185.

Celermajer, D., Schlosberg, D., Rickards, L. et al. (2021). Multispecies Justice: Theories, Challenges, and a Research Agenda for Environmental Politics. *Environmental Politics*, *30*(1–2), 119–140. https://doi.org/10.1080/09644016.2020.1827608.

Chakrabarti, P. (2020). *Inscriptions of Nature: Geology and the Naturalization of Antiquity*. Johns Hopkins University Press.

Chakrabarty, D. (2009). The Climate of History: Four Theses. *Critical Inquiry*, *35*(2), 197–222. https://doi.org/10.1086/596640.

Chakrabarty, D. (2018). Anthropocene Time. *History and Theory*, *57*(1), 5–32. https://doi.org/10.1111/hith.12044.

Chakrabarty, D. (2019). The Planet: An Emergent Humanist Category. *Critical Inquiry*, *46*(1), 1–31. https://doi.org/10.1086/705298.

Chakrabarty, D. (2021). *The Climate of History in a Planetary Age*. University of Chicago Press. https://doi.org/10.7208/9780226733050.

Chen, Z., Auler, A. S., Bakalowicz, M. et al. (2017). The World Karst Aquifer Mapping Project: Concept, Mapping Procedure and Map of Europe. *Hydrogeology Journal*, *25*(3), 771–785. https://doi.org/10.1007/s10040-016-1519-3.

Chik, H. (2021, August 24). China Eyes Mega Spacecraft Spanning Miles in Crewed Mission Push. *South China Morning Post*. www.scmp.com/news/china/science/article/3146224/china-eyes-ultra-large-spacecraft-spanning-miles-us23m-crewed.

Christian, D. (2011). *Maps of Time: An Introduction to Big History* (2nd ed.). University of California Press.

Clark, N. (2011). *Inhuman Nature: Sociable Life on a Dynamic Planet*. Sage.

Clark, N. (2014). Geo-Politics and the Disaster of the Anthropocene. *The Sociological Review*, *62*(S1), 19–37. https://doi.org/10.1111/1467-954X.12122.

Clark, N., & Szerszynski, B. (2021). *Planetary Social Thought: The Anthropocene Challenge to the Social Sciences*. Polity Press.

Clark, P. U., Shakun, J. D., Marcott, S. A. et al. (2016). Consequences of Twenty-First-Century Policy for Multi-Millennial Climate and Sea-Level Change. *Nature Climate Change*, *6*(4), 360–369. https://doi.org/10.1038/nclimate2923.

Clendenon, C. (2009a). *Hydromythology and the Ancient Greek World: An Earth Science Perspective Emphasizing Karst Hydrology*. Fineline Science Press.

Clendenon, C. (2009b). Karst Hydrology in Ancient Myths from Arcadia and Argolis, Greece. *Acta Carsologica*, *38*(1), Article 1. https://doi.org/10.3986/ac.v38i1.143.

Clendenon, C. (2009c). Ancient Greek Hydromyths about the Submarine Transport of Terrestrial Fresh Water through Seabeds Offshore of Karstic Regions. *Acta Carsologica*, *38*(2–3), Article 2–3. https://doi.org/10.3986/ac.v38i2-3.129.

Cockell, C. S., Bush, T., Bryce, C. et al. (2016). Habitability: A Review. *Astrobiology*, *16*(1), 89–117. https://doi.org/10.1089/ast.2015.1295.

Cohen, E. F. (2018). *The Political Value of Time: Citizenship, Duration, and Democratic Justice*. Cambridge University Press.

Cohen, J. J. (2015). *Stone: An Ecology of the Inhuman* (Illustrated ed.). University of Minnesota Press.

Cohen, S. N., Chang, A. C. Y., Boyer, H. W., & Helling, R. B. (1973). Construction of Biologically Functional Bacterial Plasmids in Vitro. *Proceedings of the National Academy of Sciences*, *70*(11), 3240–3244. https://doi.org/10.1073/pnas.70.11.3240.

Colebrook, C. (2017). We Have Always Been Post-Anthropocene: The Anthropocene Counterfactual. In R. Grusin (ed.), *Anthropocene Feminism* (pp. 1–20). University of Minnesota Press.

Cordier, J. M., Aguilar, R., Lescano, J. N. et al. (2021). A Global Assessment of Amphibian and Reptile Responses to Land-Use Changes. *Biological Conservation*, *253*, 108863. https://doi.org/10.1016/j.biocon.2020.108863.

Crary, A. (2023). The Toxic Ideology of Longtermism. *Radical Philosophy*, *214*, 49–57.

Cremer, C. Z., & Kemp, L. (2021). *Democratising Risk: In Search of a Methodology to Study Existential Risk*. SSRN [Preprint]. https://ssrn.com/abstract+3995225.

Crosby, A. W. (1972). *The Columbian Exchange: Biological and Cultural Consequences of 1492*. Greenwood.

Crouch, D. P. (1993). *Water Management in Ancient Greek Cities* (Illustrated ed.). Oxford University Press.

Currens, J. C. (2001). *Generalized Block Diagram of the Western Pennyroyal Karst* (Series XII) [Diagram]. Kentucky Geological Survey.

Cwerner, S. B. (2000). The Chronopolitan Ideal: Time, Belonging and Globalization. *Time & Society*, *9*(2–3), 331–345. https://doi.org/10.1177/0961463X00009002012.

Dainton, B. (2023). Temporal Consciousness. In E. N. Zalta & U. Nodelman (Eds.), *The Stanford Encyclopedia of Philosophy*. https://plato.stanford.edu/archives/spr2023/entries/consciousness-temporal/.

De La Cadena, M. (2015). *Earth Beings: Ecologies of Practice across Andean Worlds* (Illustrated ed.). Duke University Press. https://doi.org/10.1215/9780822375265.

De Waele, J., Gutiérrez, F., Parise, M., & Plan, L. (2011). Geomorphology and Natural Hazards in Karst Areas: A Review. *Geomorphology*, *134*(1–2), 1–8. https://doi.org/10.1016/j.geomorph.2011.08.001.

Dimock, W. C. (2008). *Through Other Continents: American Literature across Deep Time* (Hardcover ed.). Princeton University Press.

Donlan, C. J. (2005). Re-wilding North America. *Nature*, *436*(7053), Article 7053. https://doi.org/10.1038/436913a.

Donlan, C. J., Berger, J., Bock, C. E. et al. (2006). Pleistocene Rewilding: An Optimistic Agenda for Twenty-First Century Conservation. *The American Naturalist*, *168*(5), 660–681. https://doi.org/10.1086/508027.

Doocy, S., Daniels, A., Dooling, S., & Gorokhovich, Y. (2013). The Human Impact of Volcanoes: A Historical Review of Events 1900–2009 and Systematic Literature Review. *PLoS Currents*, *5*. https://doi.org/10.1371/currents.dis.841859091a706efebf8a30f4ed7a1901.

Du Cann, C. (2021, April 20). *Relearning the Language of a Lost World*. NOEMA. www.noemamag.com/relearning-the-language-of-a-lost-world.

Dupouey, J.-L., Dambrine, E., Laffite, J.-D., & Moares, C. (2002). Irreversible Impact of Past Land Use on Forest Soils and Biodiversity. *Ecology*, *83*(11), 2978–2984. https://doi.org/10.2307/3071833.

Dutch, S. (2020, May 24). *Geology and Election 2000*. Steve Durch Personal Website. https://stevedutch.net/research/elec2000/geolelec2000.htm.

Dyer, J. F., Bailey, C., & Tran, N. V. (2008). Ownership Characteristics of Heir Property in a Black Belt County: A Quantitative Approach. *Journal of Rural Social Sciences*, *24*(2), 192–217.

Ehrlich, H. L. (1996). How Microbes Influence Mineral Growth and Dissolution. *Chemical Geology*, *132*(1–4), 5–9. https://doi.org/10.1016/S0009-2541(96)00035-6.

Ellis, E., Maslin, M., Boivin, N., & Bauer, A. (2016). Involve Social Scientists in Defining the Anthropocene. *Nature*, *540*(7632), Article 7632. https://doi.org/10.1038/540192a.

English, B. D. (2020). *Civil Wars, Civil Beings, and Civil Rights in Alabama's Black Belt: A History of Perry County* (1st ed.). University of Alabama Press.

Fasullo, J. T., Tomas, R., Stevenson, S. et al. (2017). The Amplifying Influence of Increased Ocean Stratification on a Future Year without a Summer. *Nature Communications*, *8*(1), 1–10. https://doi.org/10.1038/s41467-017-01302-z.

Ferretti, S. (Ed.) (2021). *Space Capacity Building in the XXI Century*. Springer International. https://doi.org/10.1007/978-3-030-21938-3.

Foley, T. J. (2021). Waiting for Waste: Nuclear Imagination and the Politics of Distant Futures in Finland. *Energy Research & Social Science*, *72*, 101867. https://doi.org/10.1016/j.erss.2020.101867.

Frankenfeld, P. J. (1992). Technological Citizenship: A Normative Framework for Risk Studies. *Science, Technology, & Human Values*, *17*(4), 459–484.

Funtowicz, S., & Ravetz, J. R. (2001). Peer Review and Quality Control. *International Encyclopaedia of the Social and Behavioural Sciences*, 11179–11183. www.sciencedirect.com/science/article/abs/pii/B0080430767031971

Future Earth. (2020). *Our Future on Earth 2020: Science Insights into our Planet and Society*. www.futureearth.org/publications/our-future-on-earth.

Gadamer, H.-G. (2016). The History of the Universe and the Historicity of Human Beings. In P. Vandervelde & A. Iver (Eds.), *Hermeneutics between History and Philosophy: The Selected Writings of Hans-Georg Gadamer* (Vol. 1, pp. 25–41). Bloomsbury Academic. (Original work published 1988).

Galaz, V. (2019). Time and Politics in the Anthropocene: Too Fast, Too Slow. In F. Biermann & E. Lövbrand (Eds.), *Anthropocene Encounters: New Directions in Green Political Thinking* (pp. 109–127). Cambridge University Press.

Ganopolski, A., & Brovkin, V. (2017). Simulation of Climate, Ice Sheets and CO_2 Evolution during the Last Four Glacial Cycles with an Earth System Model of Intermediate Complexity. *Climate of the Past*, *13*(12), 1695–1716. https://doi.org/10.5194/cp-13-1695-2017.

García, A. M., & Gaviro, A. B. (2017). The Hydromythology and the Legend from Natural Events. *CTS: Revista Iberoamericana de Ciencia, Tecnología y Sociedad*, *12*(35), 183–199.

Gibson, J. S. (1941). The Alabama Black Belt: Its Geographic Status. *Economic Geography*, *17*(1), 1–23. https://doi.org/10.2307/141741.

Gilbert, S. F. (2014). A Holobiont Birth Narrative: The Epigenetic Transmission of the Human Microbiome. *Frontiers in Genetics*, *5*, 282. https://doi.org/10.3389/fgene.2014.00282.

Ginn, F., Bastian, M., Farrier, D., & Kidwell, J. (2018). Introduction: Unexpected Encounters with Deep Time. *Environmental Humanities*, *10*(1), 213–225. https://doi.org/10.1215/22011919-4385534.

Goetz, K. H. (Ed.) (2019). *The Oxford Handbook of Time and Politics*. Oxford University Press. https://doi.org/10.1093/oxfordhb/9780190862084.001.0001.

Goldscheider, N., Chen, Z., Auler, A. S. et al. (2020). Global Distribution of Carbonate Rocks and Karst Water Resources. *Hydrogeology Journal*, *28*(5), 1661–1677. https://doi.org/10.1007/s10040-020-02139-5.

Gombosi, T. I., Baker, D. N., Balogh, A. et al. (2017). Anthropogenic Space Weather. *Space Science Reviews*, *212*(3), 985–1039. https://doi.org/10.1007/s11214-017-0357-5.

González-Ricoy, I., & Gosseries, A. (Eds.) (2016). *Institutions for Future Generations*. Oxford University Press.

Gordon, H. (2021). *Notes from Deep Time: A Journey through Our Past and Future Worlds*. Profile Books.

Goudie, A. (2020). The Human Impact in Geomorphology – 50 Years of Change. *Geomorphology*, *366*, 106601. https://doi.org/10.1016/j.geomorph.2018.12.002.

Gould, S. J. (1987). *Time's Arrow, Time's Cycle: Myth and Metaphor in the Discovery of Geological Time*. Harvard University Press.

Graeber, D., & Wengrow, D. (2021). *The Dawn of Everything: A New History of Humanity*. Penguin UK.

Grajewski, B., Whelan, E. A., Lawson, C. C. et al. (2015). Miscarriage among Flight Attendants. *Epidemiology*, *26*(2), 192–203. https://doi.org/10.1097/EDE.0000000000000225.

Green, J., Airapetian, V., Bamford, R. et al. (2021). Interdisciplinary Research in Terraforming Mars: State of the Profession and Programmatics. *Bulletin of the American Astronomical Society*, *53*(4), 1–8. https://doi.org/10.3847/25c2cfeb.0408ed8f.

Guedes, J. A. d'Alpoim, Crabtree, S. A., Bocinsky, R. K., & Kohler, T. A. (2016). Twenty-First Century Approaches to Ancient Problems: Climate and Society. *Proceedings of the National Academy of Sciences*, *113*(51), 14483–14491. https://doi.org/10.1073/pnas.1616188113.

Guglielmi, A. V., & Zotov, O. D. (2007). The Human Impact on the Pc1 Wave Activity. *Journal of Atmospheric and Solar-Terrestrial Physics*, *69*(14), 1753–1758. https://doi.org/10.1016/j.jastp.2007.01.017.

Gunarti. (2021). *Gunarti: Letter from the Kendeng Mountains | Heinrich Böll Foundation | Southeast Asia Regional Office*. Heinrich-Böll-Stiftung. https://th.boell.org/en/2021/06/16/gunarti-letter-kendeng-mountains.

Hamacher, D. (2022). *The First Astronomers: How Indigenous Elders Read The Stars*. Allen & Unwin.

Hambrecht, G., Anderung, C., Brewington, S. et al. (2020). Archaeological Sites as Distributed Long-Term Observing Networks of the Past (DONOP). *Quaternary International*, *549*, 218–226. https://doi.org/10.1016/j.quaint.2018.04.016.

Hamilton, C. (2016). The Anthropocene as Rupture. *The Anthropocene Review*, *3*(2), 93–106. https://doi.org/10.1177/2053019616634741.

Hamilton, C., Bonneuil, C., & Gemenne, F. (2015). Thinking the Anthropocene. In C. Hamilton, C. Bonneuil, & F. Gemenne (Eds.), *The Anthropocene and the Global Environmental Crisis: Rethinking Modernity in a New Epoch* (pp. 1–13). Routledge.

Hanusch, F. (2018). *Democracy and Climate Change*. Routledge. https://doi.org/10.4324/9781315228983.

Hanusch, F., & Biermann, F. (2020). Deep-Time Organizations: Learning Institutional Longevity from History. *The Anthropocene Review*, *7*(1), 19–41. https://doi.org/10.1177/2053019619886670.

Hanusch, F., & Meisch, S. (2022). The Temporal Cleavage: The Case of Populist Retrotopia vs. Climate Emergency. *Environmental Politics*, 31(5), 883–903. https://doi.org/10.1080/09644016.2022.2044691.

Hanusch, F., Leggewie, C., & Meyer, E. (2021). *Planetar denken: Ein Einstieg*. Transcript. https://doi.org/10.1515/9783839453834.

Haraway, D. (2015). Anthropocene, Capitalocene, Plantationocene, Chthulucene: Making Kin. *Environmental Humanities*, *6*(1), 159–165. https://doi.org/10.1215/22011919-3615934.

Haraway, D. (2016). *Staying with the Trouble: Making Kin in the Chthulucene*. Duke University Press.

Hartmann, A., Goldscheider, N., Wagener, T., Lange, J., & Weiler, M. (2014). Karst Water Resources in a Changing World: Review of Hydrological Modeling Approaches. *Reviews of Geophysics*, *52*(3), 218–242. https://doi.org/10.1002/2013RG000443.

Hecht, G. (2018). Interscalar Vehicles for an African Anthropocene: On Waste, Temporality, and Violence. *Cultural Anthropology*, *33*(1), Article 1. https://doi.org/10.14506/ca33.1.05.

Hembry, D. H., & Weber, M. G. (2020). Ecological Interactions and Macroevolution: A New Field with Old Roots. *Annual Review of Ecology, Evolution, and Systematics*, *51*(1), 215–243. https://doi.org/10.1146/annurev-ecolsys-011720-121505.

Herridge, V. (2021). Before Making a Mammoth, Ask the Public. *Nature*, *598*(7881), 387. https://doi.org/10.1038/d41586-021-02844-5.

Horn, E. (2017). Jenseits der Kindeskinder. *Merkur*, *71*(814), 5–17.

Humphreys, S. (2022). Against Future Generations. *European Journal of International Law*, *33*(4), 1061–1092.

Hutton, M. D. J. (2010). *Theory of the Earth (1788)*. CreateSpace Independent Publishing Platform. (Original work published 1788).

Ialenti, V. (2020). *Deep Time Reckoning: How Future Thinking Can Help Earth Now*. MIT Press.

Ikejiri, T., Ebersole, J. A., Blewitt, H. L., & Ebersole, S. M. (2013). An Overview of Late Cretaceous Vertebrates from Alabama. *Bulletin of the Alabama Museum of Natural History*, *31*(1), 46–71.

International Union of Speleology. (2021). *International Year of Caves and Karst – IYCK*. https://iyck2021.org/index.php/international-year-of-caves-and-karst/.

Irvine, R. D. G. (2020). *An Anthropology of Deep Time: Geological Temporality and Social Life*. Cambridge University Press. https://doi.org/10.1017/9781108867450.

Jacobson-Tepfer, E. (2020). *The Anatomy of Deep Time: Rock Art and Landscape in the Altai Mountains of Mongolia*. Cambridge University Press. https://doi.org/10.1017/9781108855518.

James, W. (1893). *The Principles of Psychology*. H. Holt and Company.

Jarvenpaa, S. L., & Essén, A. (2023). Data Sustainability: Data Governance in Data Infrastructures across Technological and Human Generations. *Information and Organization*, 33(1), 100449.

Jasanoff, S., & Hurlbut, J. B. (2018). A Global Observatory for Gene Editing. *Nature*, *555*(7697), 435–437. https://doi.org/10.1038/d41586-018-03270-w.

Jefferson, T. (1958). Thomas Jefferson to James Madison. In J. P. Boyd (Ed.), *The Papers of Thomas Jefferson: 27 March to 30 November 1789 with*

Supplement, 19 October 1772 to 7 February 1790 (Vol. 15, pp. 392–398). Princeton University Press. https://jeffersonpapers.princeton.edu/selected-documents/thomas-jefferson-james-madison (Original work published 1789).

Johnson, C. N. (2009). Ecological Consequences of Late Quaternary Extinctions of Megafauna. *Proceedings of the Royal Society B: Biological Sciences*, *276*(1667), 2509–2519. https://doi.org/10.1098/rspb.2008.1921.

Johnson, S. S., Hebsgaard, M. B., Christensen, T. R. et al. (2007). Ancient Bacteria Show Evidence of DNA Repair. *Proceedings of the National Academy of Sciences*, *104*(36), 14401–14405. https://doi.org/10.1073/pnas.0706787104.

Kelly, E. R. (1882). *The Alternative: A Study in Psychology*. Macmillan.

Knipp, D. J., Ramsay, A. C., Beard, E. D. et al. (2016). The May 1967 Great Storm and Radio Disruption Event: Extreme Space Weather and Extraordinary Responses. *Space Weather*, *14*(9), 614–633. https://doi.org/10.1002/2016SW001423.

Koch, A., Brierley, C., Maslin, M. M., & Lewis, S. L. (2019). Earth System Impacts of the European Arrival and Great Dying in the Americas after 1492. *Quaternary Science Reviews*, *207*, 13–36. https://doi.org/10.1016/j.quascirev.2018.12.004.

Köhler, L. (2017). *Die Repräsentation von Non-Voice-Partys in Demokratien*. Springer VS. https://doi.org/10.1007/978-3-658-16700-4.

Konradus, D. (2021). Karst Ecosystems in the Vortex of Capital: A Paradigmatic Study of the Commune Link Law Politics. *Proceedings of the 2nd International Conference on Law Reform (INCLAR 2021)*, *590*, 1–6. https://doi.org/10.2991/assehr.k.211102.157.

Korver, A. P. E. (1976). The Samin Movement and Millenarism. *Bijdragen Tot de Taal-, Land- En Volkenkunde*, *132*(2), 249–266. https://doi.org/10.1163/22134379-90002642.

Koselleck, R. (2004). *Futures Past: On the Semantics of Historical Time* (K. Tribe, Trans.). Columbia University Press. (Original work published 1979).

Kutterolf, S., Jegen, M., Mitrovica, J. X. et al. (2013). A Detection of Milankovitch Frequencies in Global Volcanic Activity. *Geology*, *41*(2), 227–230. https://doi.org/10.1130/G33419.1.

Landa, M. de. (2000). *A Thousand Years of Nonlinear History* (4th printing ed.). Zone Books.

Langmuir, C. H., & Broecker, W. (2012). *How to Build a Habitable Planet: The Story of Earth from the Big Bang to Humankind* (Rev. and exp. ed.). Princeton University Press.

Lanzerotti, L. J. (2017). Space Weather: Historical and Contemporary Perspectives. *Space Science Reviews*, *212*(3), 1253–1270. https://doi.org/10.1007/s11214-017-0408-y.

Leib, K. (2015). State Sovereignty in Space: Current Models and Possible Futures. *Astropolitics*, *13*(1), 1–24. https://doi.org/10.1080/14777622.2015.1015112.

Lewis, S., & Maslin, M. A. (2018). *The Human Planet: How We Created the Anthropocene*. Pelican.

Lincoln, A. (1863). *Gettysburg Address – "Nicolay Copy"* [Webpage]. www.loc.gov/exhibits/gettysburg-address/ext/trans-nicolay-copy.html.

Linebaugh, P. (2008). *The Magna Carta Manifesto: Liberties and Commons for All*. University of California Press. http://ebookcentral.proquest.com/lib/unigiessen/detail.action?docID=345552.

Lingam, M., & Loeb, A. (2017). Impact and Mitigation Strategy for Future Solar Flares. *ArXiv:1709.05348 [Astro-Ph]*, 1–7. https://doi.org/10.48550/arXiv.1709.05348.

Losch, A. (2019). The Need of an Ethics of Planetary Sustainability. *International Journal of Astrobiology*, *18*(3), 259–266. https://doi.org/10.1017/S1473550417000490.

Lucero, L. J., & Gonzalez Cruz, J. (2020). Reconceptualizing Urbanism: Insights from Maya Cosmology. *Frontiers in Sustainable Cities*, *2*(1), 1–15. https://doi.org/10.3389/frsc.2020.00001.

Luhmann, N. (1976). The Future cannot Begin: Temporal Structures in Modern Society. *Social Research*, *43*(1), 130–152.

Macfarlane, R. (2019). *Underland: A Deep Time Journey*. W. W. Norton.

Machado, I. F., & Figueirôa, S. (2022). Mining History of Brazil: A Summary. *Mineral Economics*, *35*(2), 253–265. https://doi.org/10.1007/s13563-021-00293-0.

Macias-Fauria, M., Jepson, P., Zimov, N., & Malhi, Y. (2020). Pleistocene Arctic Megafaunal Ecological Engineering as a Natural Climate Solution? *Philosophical Transactions of the Royal Society B: Biological Sciences*, *375*(1794), 20190122. https://doi.org/10.1098/rstb.2019.0122.

Maliki, M. (2019). *Local/Global Disruption: The Response of the Samin Movement to Modernity – ProQuest* [Degree of Doctor of Philosophy, Charles Darwin University (Australia)]. https://doi.org/10.25913/5ed9cf66129d3.

Mann, C. C. (2005). *1491: New Revelations of the Americas before Columbus* (Illustrated ed.). Alfred A. Knopf.

Mann, C. C. (2011). *1493: Uncovering the New World Columbus Created*. Alfred A. Knopf.

Margulis, L., & Sagan, D. (2000). *What Is Life?* University of California Press.

Mathews, A. S. (2020). Anthropology and the Anthropocene: Criticisms, Experiments, and Collaborations. *Annual Review of Anthropology*, *49*(1), 67–82. https://doi.org/10.1146/annurev-anthro-102218-011317.

May, S. K., & Tacon, P. S. C. (2014). Kakadu National Park: Rock Art. In C. Smith (Ed.), *Encyclopedia of Global Archaeology* (pp. 4235–4240). Springer. https://doi.org/10.1007/978-1-4419-0465-2_2241.

Mayor, A. (2005). Geomythology. In R. C. Selley, L. R. M. Cocks, & I. R. Plimer (Eds.), *Encyclopedia of Geology* (Vol. 3, pp. 644–647). Academic Press. https://doi.org/10.1016/B978-0-08-102908-4.00366-0.

McClain, C. (2012, June 27). How Presidential Elections Are Impacted By a 100 Million Year Old Coastline | Deep Sea News [Marine Science Communication]. *Deep Sea News*. www.deepseanews.com/2012/06/how-presidential-elections-are-impacted-by-a-100-million-year-old-coastline/.

McPhee, J. (1981). *Basin and Range*. Farrar, Straus and Giroux (FSG).

McTaggart, J. E. (1908). I. – The Unreality of Time. *Mind*, *XVII*(4), 457–474. https://doi.org/10.1093/mind/XVII.4.457.

Meschede, M., & Warr, L. N. (2019). Germany during the Glacial Periods. In M. Meschede & L. N. Warr (Eds.), *The Geology of Germany* (pp. 259–282). Springer International. https://doi.org/10.1007/978-3-319-76102-2_16.

Miklós, R., Lénárt, L., Darabos, E. et al. (2020). Karst Water Resources and Their Complex Utilization in the Bükk Mountains, Northeast Hungary: An Assessment from a Regional Hydrogeological Perspective. *Hydrogeology Journal*, *28*(6), 2159–2172. https://doi.org/10.1007/s10040-020-02168-0.

Miller, R. (2018). *Transforming the Future*. Routledge.

Monda, E. (2018). Social Futuring – In the Context of Futures Studies. *Society and Economy*, *40*(s1), 77–109. https://doi.org/10.1556/204.2018.40.s1.5.

Montgomery, D. R. (2012). *The Rocks Don't Lie: A Geologist Investigates Noah's Flood*. W. W. Norton.

Montgomery, K. (2003). Siccar Point and Teaching the History of Geology. *Journal of Geoscience Education*, *51*(5), 500–505. https://doi.org/10.5408/1089-9995-51.5.500.

Morán, A., & Ross, M. (2021). Can Deliberation Overcome Its Extractivist Tendencies? *Deliberative Democracy Digest*. www.publicdeliberation.net/can-deliberation-overcome-its-extractivist-tendencies/.

Murphy, K. M. (2023). A Timeful Theory of Knowledge: Thunderstorms, Dams, and the Disclosure of Planetary History. *Angelaki*, *28*(1), 87–98.

Musk, E. [@elonmusk]. (2020, October 7). *@skorusARK Marginal Cost of Starship Mass to Orbit Should Be Well under $100/kg. Fully Burdened Cost*

Depends on Flight Rate. [Tweet]. Twitter. https://twitter.com/elonmusk/status/1313858597428826120.

Nelson, M. (2017). *Pushing our Limits: Insights from Biosphere 2.* University of Arizona Press.

Nelson, M. (2018). Some Ecological and Human Lessons of Biosphere 2. *European Journal of Ecology*, *4*(1), 50–55. https://doi.org/10.2478/eje-2018-0006.

Newman, C. J., & Williamson, M. (2018). Space Sustainability: Reframing the Debate. *Space Policy*, *46*, 30–37. https://doi.org/10.1016/j.spacepol.2018.03.001.

Nixon, R. (2011). *Slow Violence and the Environmentalism of the Poor.* Harvard University Press.

Northcott, M. (2015). Eschatology in the Anthropocene: From the Chronos of Deep Time to the Kairos of the Age of Humans. In C. Hamilton, C. Bonneuil, & F. Gemenne (Eds.), *The Anthropocene and the Global Environmental Crisis: Rethinking Modernity in a New Epoch* (pp. 100–111). Routledge.

Nunn, P. (2018). *The Edge of Memory: Ancient Knowledge, Oral Tradition and the Post-Glacial World.* Bloomsbury Sigma.

Nunn, P. (2021). *Worlds in Shadow: Submerged Lands in Science, Memory and Myth.* Bloomsbury Sigma.

Oeschger, H. (1985). The Contribution of Ice Core Studies to the Understanding of Environmental Processes. In W. Dansgaard, C. C. Langway, H. Oeschger (eds.), *Greenland Ice Core: Geophysics, Geochemistry, and the Environment* (pp. 9–17). American Geophysical Union (AGU). https://doi.org/10.1029/GM033p0009.

Onac, B., & van Beynen, P. (2020). Caves and Karst. In D. Alderton & S. A. Elias (Eds.), *Encyclopedia of Geology* (2nd ed., Vol. 6, pp. 495–509). Academic Press. https://digitalcommons.usf.edu/geo_facpub/2296.

Ord, T. (2020). *The Precipice: Existential Risk and the Future of Humanity* (1st ed.). Hachette Books.

Oughton, E. J., Skelton, A., Horne, R. B., Thomson, A. W. P., & Gaunt, C. T. (2017). Quantifying the Daily Economic Impact of Extreme Space Weather Due to Failure in Electricity Transmission Infrastructure. *Space Weather*, *15*(1), 65–83. https://doi.org/10.1002/2016SW001491.

Pahl, S., Sheppard, S., Boomsma, C., & Groves, C. (2014). Perceptions of Time in Relation to Climate Change. *WIREs Climate Change*, *5*(3), 375–388. https://doi.org/10.1002/wcc.272.

Paleari, C. I., Mekhaldi, F., Adolphi, F. et al. (2022). Cosmogenic Radionuclides Reveal an Extreme Solar Particle Storm near a Solar Minimum 9125 Years

BP. *Nature Communications*, *13*(1), 214. https://doi.org/10.1038/s41467-021-27891-4.

Palmer, C., McShane, K., & Sandler, R. (2014). Environmental Ethics. *Annual Review of Environment and Resources*, *39*(1), 419–442. https://doi.org/10.1146/annurev-environ-121112-094434.

Palmer, J. (2020). The New Science of Volcanoes Harnesses AI, Satellites and Gas Sensors to Forecast Eruptions. *Nature*, *581*(7808), 256–259. https://doi.org/10.1038/d41586-020-01445-y.

Pelzer, P., Hildingsson, R., Herrström, A., & Stripple, J. (2021). Planning for 1000 Years: The Råängen Experiment. *Urban Planning*, *6*(1), 249–262. https://doi.org/10.17645/up.v6i1.3534.

Playfair, J. (1805). Biographical Account of the Late Dr James Hutton, F. R. S. Edin. *Transactions of The Royal Society of Edinburgh*, *5*(3), 39–99. https://doi.org/10.1017/S0263593300090039.

Polak, F. (1973). *The Image of the Future*. Elsevier.

Poli, R. (2017). *Introduction to Anticipation Studies* (Vol. 1). Springer International. https://doi.org/10.1007/978-3-319-63023-6.

Popov, I. (2020). The Current State of Pleistocene Park, Russia (An Experiment in the Restoration of Megafauna in a Boreal Environment). *The Holocene*, *30*(10), 1471–1473. https://doi.org/10.1177/0959683620932975.

Putri, P. S. (2017). The Meaning Making of an Environmental Movement: A Perspective on Sedulur Sikep's Narrative in Anti-Cement Movement. *Power Conflict Democracy Journal*, *5*(2), 297–321. https://doi.org/10.22146/pcd.30471.

Rand, L. R., Anker, P., Fritz, D., Leigh, L., & Rosenheim, S. (2021, October 9). Biosphere 2: Why an Eccentric Ecological Experiment Still Matters 25 Years Later [Digital Magazine Produced By Graduate Students at the Center for Culture, History, and Environment (CHE)]. *Edge Effects*. https://edgeeffects.net/biosphere-2/ (Original work published 2016).

Randall, A. (1988). What Mainstream Economists Have to Say about the Value of Biodiversity. In National Academy of Sciences (Ed.), *Biodiversity* (pp. 217–224). The National Academies Press. https://doi.org/10.17226/989.

Ray, C. (2019). Sacred Wells across the Longue Durée. In C. Ray & M. Fernández-Götz (Eds.), *Historical Ecologies, Heterarchies and Transtemporal Landscapes* (pp. 265–286). Routledge.

Reinert, H. (2016). About a Stone: Some Notes on Geologic Conviviality. *Environmental Humanities*, *8*(1), 95–117. https://doi.org/10.1215/22011919-3527740.

Reisch, L. A. (2015). *Time Policies for a Sustainable Society*. Springer International. https://doi.org/10.1007/978-3-319-15198-4.

Renn, O. (2008). *Risk Governance: Coping with Uncertainty in a Complex World*. Routledge.

Riley, P. (2012). On the Probability of Occurrence of Extreme Space Weather Events. *Space Weather, 10*(2), S02012. https://doi.org/10.1029/2011SW 000734.

Riley, P., Baker, D., Liu, Y. D. et al. (2018). Extreme Space Weather Events: From Cradle to Grave. *Space Science Reviews, 214*(1), 1–24. https://doi.org/10.1007/s11214-017-0456-3.

Robinson, N. A. (2020). The Charter of the Forest: Evolving Human Rights in Nature. In R.-L. Eisma-Osorio, E. A. Kirk, & J. Steinberg Albin (Eds.), *The Impact of Environmental Law: Stories of the World We Want* (pp. 54–74). Edward Elgar.https://doi.org/10.4337/9781839106934.00010.

Saltmarshe, E., & Pembroke, B. (2019). How Art and Culture Can Help Us Rethink Time. *BBC Future*. www.bbc.com/future/article/20190521-how-art-and-culture-can-help-us-rethink-time.

Saraç-Lesavre, B. (2021). Deep Time Financing? "Generational" Responsibilities and the Problem of Rendez-Vous in the U.S. Nuclear Waste Programme. *Journal of Cultural Economy, 14*(4), 435–448. https://doi.org/10.1080/17530350.2020.1818601.

Schmidt, J. J., Brown, P. G., & Orr, C. J. (2016). Ethics in the Anthropocene: A Research Agenda. *The Anthropocene Review, 3*(3), 188–200. https://doi.org/10.1177/2053019616662052.

Schrijver, C. J., Kauristie, K., Aylward, A. D. et al. (2015). Understanding Space Weather to Shield Society: A Global Road Map for 2015–2025 Commissioned by COSPAR and ILWS. *Advances in Space Research, 55* (12), 2745–2807. https://doi.org/10.1016/j.asr.2015.03.023.

Scott, J. C. (2017). *Against the Grain: A Deep History of the Earliest States*. Yale University Press.

Serres, M. (1995). *The Natural Contract* (E. MacArthur & W. Paulson, Trans.). University of Michigan Press. https://doi.org/10.3998/mpub.9725.

Servigne, P., & Stevens, R. (2020). *How Everything Can Collapse: A Manual for Our Times* (A. Brown, Trans.; 1st ed.). Polity.

Servigne, P., Stevens, R., & Chapelle, G. (2021). *Another End of the World is Possible: Living the Collapse* (G. Samuel, Trans.; 1st ed.). Polity.

Shapiro, B. (2015). *How to Clone a Mammoth: The Science of De-Extinction*. Princeton University Press.

Shapiro, B. (2021). *Life as We Made It: How 50,000 Years of Human Innovation Refined – and Redefined – Nature*. Oneworld.

Shinohara, S. (2016). History of Organizations. In A. Farazmand (Ed.), *Global Encyclopedia of Public Administration, Public Policy, and Governance*. Springer International. https://doi.org/10.1007/978-3-319-31816-5_1-1.

Siebenhüner, B., Arnold, M., Eisenack, K., & Jacob, K. (Eds.) (2013). *Long-Term Governance for Social-Ecological Change* (1st ed.). Routledge. https://doi.org/10.4324/9780203556160.

Sigmundsson, F., Pinel, V., Lund, B. et al. (2010). Climate Effects on Volcanism: Influence on Magmatic Systems of Loading and Unloading from Ice Mass Variations, with Examples from Iceland. *Philosophical Transactions of the Royal Society A: Mathematical, Physical and Engineering Sciences*, *368*(1919), 2519–2534. https://doi.org/10.1098/rsta.2010.0042.

Silkenat, D. (2022). *Scars on the Land: An Environmental History of Slavery in the American South*. Oxford University Press.

Smail, D. L. (2021). Foreword. In G. Mackenthun and C. Mucher (eds.), *Decolonizing "Prehistory": Deep Time and Indigenous Knowledges in North America* (pp. 8–12). The University of Arizona Press. https://openresearchlibrary.org/viewer/8fe5ddc8-e7fc-4358-9168-bfa2e38e8e26

Smith, G. (2021). *Can Democracy Safeguard the Future?* Polity Press.

Sörlin, S., & Isberg, E. (2021). Synchronizing Earthly Timescales: Ice, Pollen, and the Making of Proto-Anthropocene Knowledge in the North Atlantic Region. *Annals of the American Association of Geographers*, *111*(3), 717–728. https://doi.org/10.1080/24694452.2020.1823809.

SpaceX Starlink. (2020, October 28). *Starlink Beta Terms of Service* [Reddit Post]. R/Starlink. www.reddit.com/r/Starlink/comments/jjti2k/starlink_beta_terms_of_service/.

Spate, A., & Baker, A. (2018). Karst Values of Kosciuszko National Park A Review of Values and of Recent Research. *Proceedings of the Linnean Society of New South Wales*, *140*, 253–264.

Sreekumar, K., & Hassan, G. (2020). A Study of Artificial Islands. *Journal CPPR ISSUE BRIEF, Published by Centre for Public Policy Research (CPPR)*, 1–4.

Srinivasan, K., & Kasturirangan, R. (2016). Political Ecology, Development, and Human Exceptionalism. *Geoforum*, *75*, 125–128. https://doi.org/10.1016/j.geoforum.2016.07.011.

Steffen, W., Broadgate, W., Deutsch, L., Gaffney, O., & Ludwig, C. (2015). The Trajectory of the Anthropocene: The Great Acceleration. *The Anthropocene Review*, *2*(1), 81–98. https://doi.org/10.1177/2053019614564785.

Stevanović, Z. (2019). Karst Waters in Potable Water Supply: A Global Scale Overview. *Environmental Earth Sciences*, *78*(23), 1–12. https://doi.org/10.1007/s12665-019-8670-9.

Stokes, T., Griffiths, P., & Ramsey, C. (2010). Karst Geomorphology, Hydrology, and Management. In R. Pike, D. Moore, K. Bladon, T. Redding, & R. Winkler (Eds.), *Compendium of Forest Hydrology and Geomorphology in British Columbia* (Vol. 1, pp. 373–400). Government Publication Services of British Columbia. www.for.gov.bc.ca/hfd/pubs/Docs/Lmh/Lmh66.htm.

Suchantke, I., Letizia, F., Braun, V., & Krag, H. (2020). Space Sustainability in Martian Orbits – First Insights in a Technical and Regulatory Analysis. *Journal of Space Safety Engineering*, *7*(3), 439–446. https://doi.org/10.1016/j.jsse.2020.07.003.

Sumarlan, Y., & Rumpia, J. R. (2021). *The Wong Sikep or Sedulur Sikep Movement of Central Java's Longue Durée in Its Paradoxical Nature: A Portent Form of Non-Violent Human Rights Struggles through Different Names*. Heinrich-Böll-Stiftung Southeast Asia. https://th.boell.org/en/2021/08/02/samin-sikep-central-java-pdf.

Svenning, J.-C., Pedersen, P. B. M., Donlan, C. J. et al. (2015). Science for a Wilder Anthropocene: Synthesis and Future Directions for Trophic Rewilding Research. *Proceedings of the National Academy of Sciences*, *113*(4), 898–906. https://doi.org/10.1073/pnas.1502556112.

Swinburne, R. (1990). Tensed Facts. *American Philosophical Quarterly*, *27*(2), 117–130.

Szántó, Z. O. (2018). Social Futuring – An Analytical Conceptual Framework. *Society and Economy*, *40*(s1), 5–20. https://doi.org/10.1556/204.2018.40.S1.2.

Szerszynski, B. (2017). The Anthropocene Monument: On Relating Geological and Human Time. *European Journal of Social Theory*, 20(1), 111–131.

Szerszynski, B. (2019). How the Earth Remembers and Forgets. In A. Bobbette & A. Donovan (Eds.), *Political Geology* (pp. 219–236). Springer International. https://doi.org/10.1007/978-3-319-98189-5_8.

Szerszynski, B. (2020). The Watchman's Part: Earth Time, Human Time, and the "World Scientists' Warning to Humanity." *Ecocene: Cappadocia Journal of Environmental Humanities, Cappadocia University*, *1*(1), 92–100. https://doi.org/10.46863/ecocene.39.

Táíwò, O. O. (2022). *Reconsidering Reparations* (1st ed.). Oxford University Press. https://doi.org/10.1093/oso/9780197508893.001.0001.

Talento, S., & Ganopolski, A. (2021). Reduced-Complexity Model for the Impact of Anthropogenic CO_2 Emissions on Future Glacial Cycles. *Earth*

System Dynamics, *12*(4), 1275–1293. https://doi.org/10.5194/esd-12-1275-2021.

Thompson, D. F. (2005). Democracy in Time: Popular Sovereignty and Temporal Representation. *Constellations*, *12*(2), 245–261. https://doi.org/10.1111/j.1351-0487.2005.00414.x.

Thompson, D. F. (2010). Representing Future Generations: Political Presentism and Democratic Trusteeship. *Critical Review of International Social and Political Philosophy*, *13*(1), 17–37. https://doi.org/10.1080/13698230903326232.

Tooze, A. (2022, October 22). Welcome to the World of the Polycrisis. *The Financial Times*. www.ft.com/content/498398e7-11b1-494b-9cd3-6d669dc3de33.

Torres, P. (2021). *Were the Great Tragedies of History "Mere Ripples"? The Case against Longtermism*. www.xriskology.com/_files/ugd/d9aaad_89094654cf0945738f5633b5d46653fd.pdf

Toulmin, S., & Goodfield, J. (1982). *The Discovery of Time*. University of Chicago Press.

Trauth, K. M., Hora, S. C., & Guzowski, R. V. (1993). *Expert Judgment on Markers to Deter Inadvertent Human Intrusion into the Waste Isolation Pilot Plant* (SAND-92-1382). Sandia National Labs, Albuquerque, NM. https://doi.org/10.2172/10117359.

Trofimova, E. (2018). UNESCO World Karst Natural Heritage Sites: Geographical And Geological Review. *Geography, Environment, Sustainability*, *11*(2), Article 2. https://doi.org/10.24057/2071-9388-2018-11-2-63-72.

Trump, D. (2020, May 18). *OIL (ENERGY) IS BACK!!!!* Twitter. https://twitter.com/realdonaldtrump/status/1262397998094594050.

Underdal, A. (2010). Complexity and Challenges of Long-Term Environmental Governance. *Global Environmental Change*, *20*(3), 386–393. https://doi.org/10.1016/j.gloenvcha.2010.02.005.

Uomini, N. T. (2016). Archaeology in Karst Areas. *Zeitschrift Für Geomorphologie*, *60*(Supplementary Issue 2), 129–138. https://doi.org/10.1127/zfg_suppl/2016/00318.

Ussher, J. (1650). *Annals of the World*. http://archive.org/details/AnnalsOfTheWorld.

Van Beynen, P., & Townsend, K. (2005). A Disturbance Index for Karst Environments. *Environmental Management*, *36*(1), 101–116. https://doi.org/10.1007/s00267-004-0265-9.

van der Leeuw, S. (2020). *Social Sustainability, Past and Future: Undoing Unintended Consequences for the Earth's Survival*. Cambridge University Press. https://doi.org/10.1017/9781108595247.

Van Eijk, C. (2020). Sorry, Elon: Mars Is Not a Legal Vacuum – and It's Not Yours, Either. *Völkerrechtsblog*. https://doi.org/10.17176/20210107-183703-0.

Venditti, B. (2022, January 27). *The Cost of Space Flight before and after SpaceX*. Visual Capitalist. www.visualcapitalist.com/the-cost-of-space-flight/.

Vitaliano, D. B. (1973). *Legends of the Earth: Their Geologic Origins*. Indiana University Press.

Wadham, J. L., Hawkings, J. R., Tarasov, L. et al. (2019). Ice Sheets Matter for the Global Carbon Cycle. *Nature Communications*, *10*(1), Article 1. https://doi.org/10.1038/s41467-019-11394-4.

Walker, G. (2021). *Energy and Rhythm: Rhythmanalysis for a Low Carbon Future*. Rowman & Littlefield.

Walker [@Sara_Imari], S. I. (2022, January 6). *The First Cell Never Died, It Bifurcated in Time, Generated All Biology, All of Us, and All Technology. Life Is a Lineage of Events – Everything We Do Becomes a Part of That Lineage.* [Tweet]. Twitter. https://twitter.com/Sara_Imari/status/1479122723087347727.

Wang, C., Hazen, R. M., Cheng, Q. et al. (2021). The Deep-Time Digital Earth Program: Data-Driven Discovery in Geosciences. *National Science Review*, *8*(9), nwab027. https://doi.org/10.1093/nsr/nwab027.

Washington, B. T. (1901). Up from Slavery. Doubleday, Page.

Westengen, O. T., Jeppson, S., & Guarino, L. (2013). Global Ex-Situ Crop Diversity Conservation and the Svalbard Global Seed Vault: Assessing the Current Status. *PLoS One*, *8*(5), e64146. https://doi.org/10.1371/journal.pone.0064146.

Westermann, A., & Rohr, C. (2015). Climate and Beyond: The Production of Knowledge about the Earth as a Signpost of Social Change; An Introduction. *Historical Social Research / Historische Sozialforschung*, *40*(2), 7–21. https://doi.org/10.12759/HSR.40.2015.2.7-21.

Whatmore, S. (2006). Materialist Returns: Practising Cultural Geography in and for a More-Than-Human World. *Cultural Geographies*, *13*(4), 600–609. https://doi.org/10.1191/1474474006cgj377oa.

White, W. B., Herman, J. S., Herman, E. K., & Rutigliano, M. (2018). *Karst Groundwater Contamination and Public Health: Beyond Case Studies*. Springer. https://doi.org/10.1007/978-3-319-51070-5.

Whiteside, K. H. (2018). Future Generations and the Limits of Representation. In D. Castiglione & J. Pollak (Eds.), *Creating Political Presence: The New Politics of Democratic Representation* (pp. 204–228). University of Chicago Press.

Willeit, M., Ganopolski, A., Calov, R., & Brovkin, V. (2019). Mid-Pleistocene Transition in Glacial Cycles Explained by Declining CO_2 and Regolith Removal. *Science Advances*, *5*(4), eaav7337. https://doi.org/10.1126/sciadv.aav7337.

Williams, C., & Nield, T. (2007). Earth's Next Supercontinent. *New Scientist*, *196*(2626), 36–40. https://doi.org/10.1016/S0262-4079(07)62661-X.

Wilson, T. M., Stewart, C., Sword-Daniels, V. et al. (2012). Volcanic Ash Impacts on Critical Infrastructure. *Physics and Chemistry of the Earth, Parts A/B/C*, *45–46*, 5–23. https://doi.org/10.1016/j.pce.2011.06.006.

Witze, A. (2020). How a Small Nuclear War Would Transform the Entire Planet. *Nature*, *579*, 485–487. https://doi.org/10.1038/d41586-020-00794-y.

Wójtowicz, T., & Szocik, K. (2021). Democracy or What? Political System on the Planet Mars after Its Colonization. *Technological Forecasting and Social Change*, *166*, 120619. https://doi.org/10.1016/j.techfore.2021.120619.

Wood, D. (2018). *Deep Time, Dark Times: On Being Geologically Human*. Fordham University Press.

Yamagata, K., Nagai, K., Miyamoto, H. et al. (2019). Signs of Biological Activities of 28,000-Year-Old Mammoth Nuclei in Mouse Oocytes Visualized By Live-Cell Imaging. *Scientific Reports*, *9*, Article 1. https://doi.org/10.1038/s41598-019-40546-1.

Younos, T., Schreiber, M., & Kosič Ficco, K. (2018). Preface. In T. Younos, M. Schreiber, K. K. Ficco (eds.), *Karst Water Environment: Advances in Research, Management and Policy* (Vol. 68, pp. xi–xiv). Springer.

Yusoff, K. (2016). Anthropogenesis: Origins and Endings in the Anthropocene. *Theory, Culture & Society*, *33*(2), 3–28. https://doi.org/10.1177/0263276415581021.

Yusoff, K. (2018a). Politics of the Anthropocene: Formation of the Commons as a Geologic Process: Politics of the Anthropocene. *Antipode*, *50*(1), 255–276. https://doi.org/10.1111/anti.12334.

Yusoff, K. (2018b). *A Billion Black Anthropocenes or None*. University of Minnesota Press.

Zalasiewicz, J., & Kunkel, B. (2017, April 7). *"We Are the Meteor" | On the Media* (B. Gladstone, Interviewer) [Podcast]. www.wnyc.org/story/we-are-meteor/.

Zalasiewicz, J., Waters, C. N., Williams, M., & Summerhayes, C. P. (Eds.) (2019). *The Anthropocene as a Geological Time Unit: A Guide to the Scientific Evidence and Current Debate*. Cambridge University Press.

Zapletalová, J., Stefanova, D., Vaishar, A. et al. (2016). Social Development of Ecologically Sensitive Rural Areas: Case Studies of the Moravian Karst

(Czech Republic) and the Devetashko Plato (Bulgaria). *ПРОБЛЕМИ НА ГЕОГРАФИЯТА • Problems of Geography, 3–4*, 23.

Zen, E. (2001). What is Deep Time and Why Should Anyone Care? *Journal of Geoscience Education, 49*(1), 5–9. https://doi.org/10.5408/1089-9995-49.1.5.

Zielinski, S. (2013). Vorwort. In S. Zielinski and E. Fürlus (eds.), *Variantologie: Zur Tiefenzeit der Beziehungen von Kunst, Wissenschaft und Technik* (pp. 7–16). Kulturverlag Kadmos.

Zimov, S. A. (2005). Pleistocene Park: Return of the Mammoth's Ecosystem. *Science, 308*(5723), 796–798. https://doi.org/10.1126/science.1113442.

Ziolkowski, T. (1990). *German Romanticism and Its Institutions*. Princeton University Press. https://doi.org/10.1515/9780691225760.

Acknowledgments

I would like to express my sincere gratitude to Justus Liebig University Giessen and THE NEW INSTITUTE for their financial support, which made the publication of this Element possible.

I would particularly like to thank Jessica Mörsdorf for her valuable contributions to the manuscript, including proofreading, formatting, and assisting in the organization of the references and figures, as well as Meike Wiegand for her help in regard to the permission of figures. I would also like to thank two anonymous reviewers who provided very useful comments, which sharpened the conceptual clarity and overall argumentation of the Element. Any shortcomings or failures are solely my own responsibility.

About the Author

Frederic Hanusch is scientific manager of the Panel on Planetary Thinking at Gießen University and a fellow at THE NEW INSTITUTE in Hamburg. His research explores the interactions between democracy and planetary change. He provides analytical foundations for democratic renewal and identifies opportunities for interdisciplinary collaborations.

Cambridge Elements

Earth System Governance

Frank Biermann
Utrecht University

Frank Biermann is Research Professor of Global Sustainability Governance with the Copernicus Institute of Sustainable Development, Utrecht University, the Netherlands. He is the founding Chair of the Earth System Governance Project, a global transdisciplinary research network launched in 2009; and Editor-in-Chief of the new peer-reviewed journal *Earth System Governance* (Elsevier). In April 2018, he won a European Research Council Advanced Grant for a research program on the steering effects of the Sustainable Development Goals.

Aarti Gupta
Wageningen University

Aarti Gupta is Professor of Global Environmental Governance at Wageningen University, The Netherlands. She is Lead Faculty and a member of the Scientific Steering Committee of the Earth System Governance (ESG) Project and a Coordinating Lead Author of its 2018 Science and Implementation Plan. She is also principal investigator of the Dutch Research Council-funded TRANSGOV project on the Transformative Potential of Transparency in Climate Governance. She holds a PhD from Yale University in environmental studies.

Michael Mason
London School of Economics and Political Science

Michael Mason is a full professor in the Department of Geography and Environment at the London School of Economics and Political Science. At LSE he is also Director of the Middle East Centre and an Associate of the Grantham Institute on Climate Change and the Environment. Alongside his academic research on environmental politics and governance, he has advised various governments and international organisations on environmental policy issues, including the European Commission, ICRC, NATO, the UK Government (FCDO), and UNDP.

About the Series

Linked with the Earth System Governance Project, this exciting new series will provide concise but authoritative studies of the governance of complex socio-ecological systems, written by world-leading scholars. Highly interdisciplinary in scope, the series will address governance processes and institutions at all levels of decision-making, from local to global, within a planetary perspective that seeks to align current institutions and governance systems with the fundamental 21st Century challenges of global environmental change and earth system transformations.

Elements in this series will present cutting edge scientific research, while also seeking to contribute innovative transformative ideas towards better governance. A key aim of the series is to present policy-relevant research that is of interest to both academics and policy-makers working on earth system governance.

More information about the Earth System Governance project can be found at: www.earthsystemgovernance.org

Cambridge Elements

Earth System Governance

Elements in the Series

Remaking Political Institutions: Climate Change and Beyond
James J. Patterson

Forest Governance: Hydra or Chloris?
Bas Arts

Decarbonising Economies
Harriet Bulkeley, Johannes Stripple, Lars J. Nilsson, Bregje van Veelen,
Agni Kalfagianni, Fredric Bauer and Mariësse van Sluisveld

Changing Our Ways: Behaviour Change and the Climate Crisis
Peter Newell, Freddie Daley and Michelle Twena

The Carbon Market Challenge: Preventing Abuse Through Effective Governance
Regina Betz, Axel Michaelowa, Paula Castro, Raphaela Kotsch, Michael Mehling,
Katharina Michaelowa and Andrea Baranzini

*Addressing the Grand Challenges of Planetary Governance: The Future
of the Global Political Order*
Oran R. Young

Adaptive Governance to Manage Human Mobility and Natural Resource Stress
Saleem H. Ali, Martin Clifford, Dominic Kniveton, Caroline Zickgraf
and Sonja Ayeb-Karlsson

*The Emergence of Geoengineering: How Knowledge Networks
Form Governance Objects*
Ina Möller

The Normative Foundations of International Climate Adaptation Finance
Romain Weikmans

Just Transitions
Dimitris Stevis

A Green and Just Recovery from COVID-19?
Kyla Tienhaara, Tom Moerenhout, Vanessa Corkal, Joachim Roth, Hannah
Ascough, Jessica Herrera Betancur, Samantha Hussman, Jessica Oliver,
Kabir Shahani and Tianna Tischbein

The Politics of Deep Time
Frederic Hanusch

A full series listing is available at www.cambridge.org/EESG

Printed by Printforce, United Kingdom